SpringerBriefs in Bioengineering

For further volumes:
http://www.springer.com/series/10280

Ivan V. Maly

Systems Biomechanics
of the Cell

 Springer

Ivan V. Maly
Department of Computational Biology
School of Medicine
University of Pittsburgh
Pittsburgh, PA, USA

ISSN 2193-097X   ISSN 2193-0988 (electronic)
ISBN 978-1-4614-6882-0   ISBN 978-1-4614-6883-7 (eBook)
DOI 10.1007/978-1-4614-6883-7
Springer New York Heidelberg Dordrecht London

Library of Congress Control Number: 2013933236

Printed on acid-free paper

Springer is part of Springer Science+Business Media (www.springer.com)

Contents

Systems Biomechanics of the Cell

Introduction

Systems biology is a branch of biological sciences on the frontier of today's inquiry into the mechanisms of life. It studies nonintuitive effects of interaction of multiple elements in biological systems, aiming to complement the study of the composition of living matter (molecules, species, etc.) and to bring the mathematical methods capable of describing complex processes to bear on biological problems. The early work in systems biology, which originated in its modern form in the 1960s, was remarkably diverse (reviewed in Maly 2009). In comparison, most of the recent effort has been motivated by the data explosion in descriptive genomics and proteomics and devoted to the regulation of transcription and metabolism (see, e.g., Longabaugh and Bolouri 2006; Orth et al. 2010). These well-developed research directions address the system-level problems in the *chemistry* of the living matter. The *mechanics*, on the other hand, has so far attracted far less interest among systems biologists. It is the mechanics, however, that is responsible for the most obvious characteristics of the living matter—for the generation of its complex form and movement. As any microscopist can attest, the complex (at the same time "whimsical" and "purposeful") form and movement of living matter become unmistakable already on the level of single cells. Biomechanics of the cell therefore appears to be a worthy application for the systems method. This short monograph presents a personal view on the most elementary systems-biomechanical effects that arise on the cellular level of biological organization, and on how they can be analyzed with adequate quantitative rigor.

Structurally and mechanically, in the generic eukaryotic cell it is useful to distinguish between the cell boundary and the cell body (Harold 2001; Baluska et al. 2004). The cell boundary appears invariably to be comprised of the actin cortex underlying the plasma membrane; extraplasmatic material of varying mechanical strength is common, which in plants and fungi may take the form of a rigid cell wall. The cell body, properly so defined, is included inside the boundary and is comprised of the nucleus and the juxtanuclear complex of microtubules and membranous organelles. The degree of mechanical association between these components and therefore the structural composition of the cell body proper may vary between cell types and stages of development, but the presence of a single dominant, mechanically linked complex inside the cell boundary seems to be typical, giving rise to the cell body concept. The similarity of prokaryotic cells to the eukaryotic body-boundary organization appears increasing in the light of the recent research into the prokaryotic cytoskeleton; here we will be concerned, however, with the eukaryotic structure as the one which is currently better studied and which gives rise to the more complex forms and movements that exemplify advanced life.

It seems natural to structure the present approach to the system biomechanics of the cell by focusing on the system-level effects in the mechanics of the body, the boundary, and their interaction. Furthermore, in the first approximation it seems worthwhile exploring the effects limited to those arising from the mechanics of the body reduced to an idealized aster (or asters) or microtubules flexing elastically, and of the boundary reduced to an idealized shell, either rigid or under line tension. The interaction between the body and the boundary can be limited to simple contact; cortically anchored molecular motors exerting dissipative forces on the contacting microtubules can be considered in addition. Although this representation of the cell mechanics is extremely idealized and simplified, the presented analysis shows that the limited set of simple physical forces, when placed under the structural constraints of the generic heritable cell structure (body and boundary), can produce remarkably complex and life-like behavior. The particularly fundamental and general emergent effects that will be considered in this book are instability of symmetry (which makes the specific cases of stability nontrivial), irreversibility, and dissipative oscillations. The thesis of this short monograph is that system-level mechanical effects on the cellular level deserve further theoretical study, must be taken into account when designing and interpreting experiments, and may ultimately prove to be responsible for many of the most intriguing manifestations of life.

Instability of Symmetry

Unipolar Cell Body

In many cell types that have been studied experimentally, the prominent structural feature is the aster of microtubules, which are converging and anchored on the centrosome (Bray 2001). Besides forming the structural basis for the cell body as defined in the introduction, the aster's microtubules direct transport of organelles and secretory vesicles. This makes the position of the centrosome a reliable marker for the position of the cell body within the cell boundary as well as functionally important for various cellular activities, such as wound closure, migration, and interactions between cells of the immune system (Gotlieb et al. 1981; Kupfer et al. 1982, 1994; Kupfer and Singer 1989; Ueda et al. 1997). In particular, induction of asymmetry of the microtubule aster in some situations, or maintenance of its symmetry within the cell outline in others, has been a subject of experimental investigation (Kupfer et al. 1982; Euteneuer and Schliwa 1992; Koonce et al. 1999; Piel et al. 2000; Etienne-Manneville and Hall 2001; Yvon et al. 2002; Burakov et al. 2003; Gomes et al. 2005; Dupin et al. 2009).

In the compact immune cells, such as the lymphocytes, which have nearly spherical cell bodies, the eccentric position of the centrosome appears to be constitutive, and only its orientation to a specific side of the cell is regulated by the antigen-mediated cell–cell interaction. In the thinly spread epithelioid cells of the wound-closure experiments, which with the exception of the nucleus are nearly flat, the centrality and eccentricity of the centrosome were initially a matter of debate. It is now generally accepted that although the centrosome in these cells may not be centered with respect to the nucleus, it is centered with respect to the cell outline. The terminology adopted here makes this observation clearer: Although the orientation of the cell body may differ, it is centered in the cell boundary. The apparent fundamental difference between the centrosome positioning in flat and spherical cells would require explanation and might serve as a test of our understanding of the centrosome positioning mechanisms; however, the quantitative analysis reviewed below casts doubt on this interpretation of the experimental results. Irrespectively, positioning of the cell body and especially of the centrosome as its best-defined structural and functional marker within the cell boundary is of central importance in the biology of the cell.

The mechanism of the positioning is not well understood. It is unclear how general the mechanisms are that have been implicated, or how exactly they interact or interfere in each cell type and in each individual cell. Among the experimentally implicated mechanisms there are microtubule dynamics (Stowers et al. 1995; Lowin-Kropf et al. 1998; Faivre-Moskalenko and Dogterom 2002; Yvon et al. 2002; Burakov et al. 2003), action of cortically anchored molecular motors of the dynein type (Etienne-Manneville and Hall 2001; Burakov et al. 2003; Levy and Holzbaur 2008), movement of the entire cell body that entrains the centrosome (Arkhipov and Maly 2006a, 2008), flow of cortical actomyosin that entrains microtubules (Burakov et al. 2003; Gomes et al. 2005), and even cell population-level kinetic selection linked to the direction of transport along the microtubules (Arkhipov and Maly 2006b, 2007).

In view of the complexity of centrosome positioning and with the goal of progressing toward a generalized and integrated mechanistic understanding of it, it is imperative to study systematically the contributions and theoretical capacities of each contributing mechanism, starting from the first principles. Arguably the simplest of the contributions, and the one which is the most inseparable from the microtubule cytoskeleton itself, is the effect that the bending elasticity of microtubules must have on the positioning of the centrosome within the cell boundary.

The fundamental role of the elastic compactization of the microtubule cytoskeleton within the constraints of the cell boundary for centrosome positioning was recognized early by Holy. One version of the original theory considered the absolute energy minimum of an aster of a large finite number of evenly spaced microtubules of equal length, confined to a flat circular domain (Holy 1997). It was shown that when the microtubule length exceeds the confining radius, the centrosome would become positioned eccentrically, and the equilibrium centrosome position was computed as a function of the length and radius. An argument was made that the instability of the central position must be stronger in three dimensions than in two dimensions. The other variant of this pioneering theory (Holy et al. 1997) incorporated also stochastic changes in the lengths of individual microtubules, which might also affect centrosome positioning. The theory predicted approximate centering, when the microtubules were comparable in length with the radius of the confining boundary, and eccentric positioning of the centrosome, when the microtubules were significantly longer. The asymmetric unstrained configuration of the model aster had an impact on the results of this pioneering study. In later theories that incorporated the microtubule bending elasticity (Arkhipov and Maly 2006a; Howard 2006; Baratt et al. 2008; Kim and Maly 2009; Pinot et al. 2009), its effect was similarly compounded by other simultaneously acting mechanisms such as the stochastic microtubule assembly, cell surface dynamics, action of molecular motors, and Brownian movement. Among the cited works, the theory by Howard (2006) predicted centering of the centrosome in the cell, irrespective of the number of dimensions. In this respect it was similar to the one-dimensional theory without bending (Dogterom and Yurke 1998), whose principles proved applicable to fission yeast (Tran et al. 2001). The other models predicted an eccentric position for the centrosome in two as well as in three dimensions.

Maly and Maly (2010) examined equilibria of an idealized microtubule cytoskeleton (an aster of microtubules converging on the centrosome) as they are dictated by the microtubule bending elasticity alone within the constraint of the cell boundary. Special attention was paid to the question whether symmetry or asymmetry of the cytoskeleton is favored. In view of this goal, an idealized cytoskeleton whose unstrained configuration is symmetric was considered. The model was applied to essentially two-dimensional (flat) and to three-dimensional cytoplasmic domains, which mimic the experimentally studied cases of epithelioid and immune cells.

The analysis conducted by Maly and Maly (2010) demonstrates that the asymmetry of a microtubule cytoskeleton will arise from the instability of the symmetric equilibrium due to the microtubule bending elasticity alone. The bending elasticity of microtubules is a property of any microtubule cytoskeleton. The asymmetries observed in the experiments and in previous simulations that incorporated the bending elasticity, therefore, cannot be attributed to the other factors that also were at work: to asymmetry of the unstrained configuration of the cytoskeleton itself, to external asymmetries, to stochasticity, or to other forces. Instead, asymmetry in the light of the analysis by Maly and Maly (2010) should be viewed as inevitably arising in any confined microtubule cytoskeleton, insofar as its microtubules bend.

The formal stability analysis of the symmetric state in this model supports the cited earlier theories that predicted emergence of asymmetry. The simple model also allowed to identify the one factor in the previous multifactorial models which alone would be sufficient for such emergence, namely the microtubule bending elasticity. The analysis at the same time showed that the equality of forces from

Fig. 1 Diagram of the
model for the equilibrium
of a confined centrosomal
microtubule. Reproduced
from Maly and Maly (2010)
with permission from
Elsevier

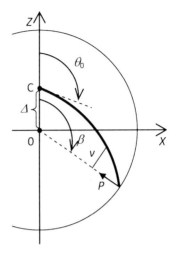

buckled microtubules on different sides of the centrosome in the symmetric case (Howard 2006) is not sufficient to predict the symmetry, because this equilibrium will be unstable.

Compared with the existing theory of confined microtubule asters shaped by microtubule bending elasticity alone (Holy 1997), the key new feature of the analysis by Maly and Maly was specification of the centered symmetric equilibrium conformation. It made possible the formal stability analysis of such a state. It also led to the recognition that the asymmetric equilibrium established in flat cells does not in general correspond to the absolute energy minimum that the original theory (Holy 1997) considered. Instead, the asymmetric equilibrium displays special stability properties and history-dependence. This accounts for the different prediction that the new model gives for the coefficient of approximate proportionality between the equilibrium distance of the centrosome from the center and the difference of the microtubule length and cell radius. Maly and Maly predict a coefficient of 1, whereas the initial slope in the energy-minimum theory was close to 1.5 (before the function plateaued with the centrosome approaching the boundary). Extending this analysis to three-dimensional cells, Maly and Maly found that the asymmetric equilibrium in this case corresponds to the absolute energy minimum, and computed the coefficient of the approximate proportionality as equal to 2. The pioneering theory (Holy 1997) put forward an argument that the instability of the centered position should be stronger in three dimensions than in two. Maly and Maly (2010) were able to demonstrate that in two dimensions, the instability is third-order, and in three dimensions it is of a special kind: infinitesimal displacements lead to a finite force.

In the Maly and Maly model, the task of finding the equilibrium conformation of the microtubule cytoskeleton is divided into finding the equilibrium forms of microtubules and finding the equilibrium position of the centrosome. Figure 1 shows the geometry of the model insofar as finding the equilibrium of a single microtubule is

Table 1 Nomenclature of the symmetry instability model

Symbol	Meaning
β	Angular coordinate of the microtubule contact with the boundary
Δ	Distance between the centrosome and the cell center
Δ_e	Equilibrium distance between the centrosome and the cell center
θ	Angle of the tangent to the microtubule
θ_0	Angle at which the microtubule is clamped at the centrosome
EI	Microtubule flexural rigidity
F	Force exerted by all microtubules on the centrosome
f	Vertical component of the force exerted by a microtubule on the centrosome
L	Microtubule length
M	Moment of the microtubule–boundary contact force
N	Number of microtubules in the cell
n_i	Fraction of microtubules in stable ($i=1$) and metastable ($i=2$) forms
P	Contact force from the cell boundary on the distal microtubule end
$p = N/(2\pi)$	Density of microtubules per unit of θ_0 in a flat cell
$p = N/(4\pi)$	Density of microtubules per unit of the solid angle Ω in a three-dimensional cell
R	Cell radius
s	Axial coordinate in a microtubule
v	Moment arm of the microtubule–boundary contact force

concerned. Table 1 lists the model parameters. The following equations specify the model for the single microtubule. They consist of the standard equilibrium beam equation, and of the boundary conditions of clamping on the centrosome and frictionless contact with the cell boundary:

$$\frac{\mathrm{d}}{\mathrm{d}s}x(s) = \sin\theta(s), \ \frac{\mathrm{d}}{\mathrm{d}s}z(s) = \cos\theta(s), \ \mathrm{d}s = \sqrt{\mathrm{d}x^2 + \mathrm{d}z^2}$$

$$\frac{\mathrm{d}}{\mathrm{d}s}\theta(s) = \frac{M(s)}{\mathrm{EI}}, \ M(s) = Pv(s), \ v(s) = -x(s)\cos\theta + z(s)\sin\theta$$

$$x(0) = 0, \ z(0) = \Delta, \ \theta(0) = \theta_0, \ x(L) = R\sin\theta, \ z(L) = R\cos\theta$$

Since

$$x(L) = x(L, \Delta, \theta_0, P, \beta), \ z(L) = z(L, \Delta, \theta_0, P, \beta)$$

the contact conditions, which are the characteristic equations for the nonlinear boundary problem, specify the unknown parameters P and β as functions

$$P = P(L, \Delta, \theta_0), \quad \beta = \beta(L, \Delta, \theta_0)$$

In the three-dimensional case there is only one stable solution, and, correspondingly, one (P, β) pair. In the flat (two-dimensional) case, the nonlinearity leads to existence of two types of stable equilibrium forms of microtubules, and accordingly there are two branches of the functions for the unknown parameters:

$$P_i = P_i(L, \Delta, \theta_0), \quad \beta_i = \beta_i(L, \Delta, \theta_0), \quad i = 1, 2$$

With the known parameters (P, β), the differential equations of bending determine the deformed shape of each microtubule and the stresses in it. In particular, the action of each microtubule on the centrosome will be known. For each microtubule, we have the component of the force

$$f(\theta_0) = -P(\theta_0)\cos\beta(\theta_0)$$

that contributes to the total force on the centrosome. In the two-dimensional case it is found as

$$F = \sum_{i=1}^{2} \int_0^\pi 2f(\theta_0)n_i(\theta_0)p\,d\theta_0$$

and in the three-dimensional case as

$$F = \int f(\theta_0)p\,d\Omega$$

The equilibrium condition for the centrosome, and for the microtubule cytoskeleton as a whole, is

$$F = 0$$

The model is solved numerically. It can be observed that when $\Delta = 0$, the equilibrium problem for the single microtubule is equivalent to the Euler problem about an elastic rod hinged on both ends. Indeed, even though the centrosomal end of the microtubule is not hinged, the moment of the force P at the centrosomal end is zero when $\Delta = 0$. Maly and Maly (2010) developed the following approximation of the Euler–Lagrange solution to the Euler problem:

$$P = \frac{\pi^2 \kappa}{L^2} \frac{1 + 0.09\varepsilon^3}{1 - \dfrac{1}{2}\varepsilon}, \quad \beta = \frac{2\sqrt{\varepsilon}}{1 - \dfrac{5}{48}\varepsilon - \dfrac{1}{53}\varepsilon^2}$$

Here $\varepsilon = (L - R)/L$. These approximations were shown to have accuracies 0.25 % and 0.05 %, respectively. The ballistic method can be used to refine the solution, and then to continue it with respect to Δ. The domain of existence of the metastable solution $(i = 2)$ in the two-dimensional case has to be mapped numerically in the space (L, θ_0, Δ). Variable step sizes need to be used to accurately account for the singularity at the boundary of this domain when continuing the solution and integrating the force.

The specific assumptions in the Maly and Maly model about the mechanical system of the microtubule aster confined within the cell boundary merit some discussion. The simple beam equation is the classical model for microtubule bending that has found both theoretical and experimental applications (see Howard 1998). The adequacy of the beam equation, which predicts the simple Euler forms of buckling, in application to intracellular microtubules is not universal in the light of observations of high-frequency buckling (Brangwynne et al. 2006). An example of more advanced models for bending of single microtubules can be found in the work of Gu et al. (2009). It may be expected that the qualitative conclusions pertaining to the properties of symmetry, stability, and reversibility that the simple model has illuminated do not depend on how simple the buckled form is. At the same time, the simple buckling forms compatible with this model are also abundant in cells, as the microphotographs in the cited papers attest.

That the microtubules are clamped rather than hinged on the centrosome may be deduced from images in the cited experimental papers. The images invariably show that even strongly bent microtubules radiate from the centrosome in all directions before bending, instead of converging on the centrosome at sharp angles in a fan-like arrangement. The situation is opposite in cell-free and centrosome-free experimental models and their numerical representation (Pinot et al. 2009) which are discussed in more detail in the section devoted to boundary dynamics. The model discussed here is concerned with the mechanical system responsible for the positioning of actual centrosomes and depends on the faithful description of the mode of microtubule clamping on them.

Maly and Maly observed that when the centrosome is in the center of the cell, the forms of a microtubule that differ in the direction of its buckling are symmetric about the axis which is the direction at which the microtubule is clamped at the centrosome, and they have the same energy. In three dimensions (in a spherical cell), an infinite number of such forms are connected by continuous rotation about this axis. In two dimensions (in a flat round cell), there are only two such forms of a microtubule, and they correspond to diametrically opposed forms of the three-dimensional case. With the displacement of the centrosome from the cell center, the described equivalency of the buckling forms of a microtubule is lost. Now the form which is convex in the direction of the centrosome displacement is bent less, and the opposite form is bent more. In three dimensions, these two diametrically opposed forms are the only remaining equilibrium forms. They lie in the plane defined by the displacement of the centrosome and by the unstrained direction of the microtubule. They are connected by a continuity of nonequilibrium forms, and there is no energy barrier between them. The microtubule therefore will adopt the lowest-energy form (the one which is convex in the direction of the centrosome displacement from the center). In two dimensions, the two equilibrium forms are connected only by higher-energy nonequilibrium forms in the plane of the cell. Therefore both of them are locally stable, and both can be occupied even after the displacement of the centrosome from the center. The higher-energy form can be called metastable, and the lower-energy form, stable.

Each equilibrium form of a microtubule is characterized by the force and torque that it exerts on the centrosome. If the total force and torque exerted on the

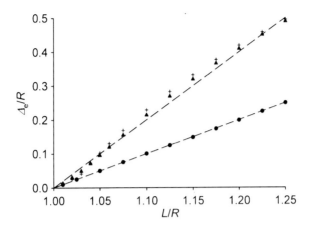

Fig. 2 Equilibrium distance of the centrosome from the cell center. Continuous three-dimensional model (*Triangles*). Discrete model with 20 microtubules emanating in the directions of the vertices of a dodecahedron (*Crosses*). Continuous two-dimensional model (positions reached spontaneously after a small perturbation of a fully symmetric cytoskeleton) (*Circles*). Slopes 1 and 2 for reference (*Dashed*). Reproduced from Maly and Maly (2010) with permission from Elsevier

centrosome by all microtubules are zero, the entire microtubule cytoskeleton will be in equilibrium inside the cell. It is such equilibrium forms of the microtubule cytoskeleton as a whole that are of interest to the theory of centrosome positioning and cell body positioning within the cell boundary. The simplest cytoskeleton is characterized by a uniform distribution of the unstressed directions at which the microtubules are clamped at the centrosome. In this case, a displacement of the centrosome from the center preserves one axis of symmetry, which coincides with the displacement. This residual symmetry makes the total torque zero and the total force collinear with the centrosome displacement.

In three dimensions, as has been mentioned, any deviation of the centrosome from the cell center specifies the direction of buckling for each microtubule, which is convex in the direction of the centrosome displacement. Thus, any such deviation generates a form of the cytoskeleton in which there is a non-zero total force on the centrosome, and the direction of this force is away from the center. Calculations show (Fig. 2) that the equilibrium is reached when the centrosome is removed from the cell center by a distance that is approximately twice as large as the difference between the microtubule length and cell radius. Comparison of the numerical results (Fig. 2) shows that the simple continuous approximation for the microtubule aster is already accurate when the number of microtubules is much lower than is typical in mammalian cells. Figure 3 shows the shape of the three-dimensional cytoskeleton at equilibrium.

The illustrated case of the spherical cell reproduces the salient features of the internal structure of the quasi-spherical lymphocytes, namely the eccentric position of the centrosome and the "combed" arrangement of the microtubules (Kuhn and Poenie 2002). This lends support to the model which is reviewed below in the section devoted to the cell boundary dynamics and which asserts that the

Fig. 3 Equilibrium
conformation of the
microtubule cytoskeleton in a
spherical cell. Sample
microtubule forms that lie in
the plane passing through the
centrosome and the cell
center are shown.
Reproduced from Maly and
Maly (2010) with permission
from Elsevier

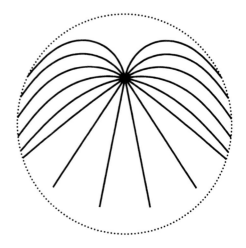

microtubule cytoskeleton in these cells adopts the minimum-energy conformation, while the antigen-mediated conjugation with another cell provides the external reference frame with respect to which the other forces may orient the constitutively asymmetric microtubule aster (Arkhipov and Maly 2006a; Baratt et al. 2008).

As an example with real numbers, consider an experiment in which an aster of $N = 20$ microtubules, each $L = 12$ μm long, is assembled inside an approximately spherical chamber of radius $R = 10$ μm, with a bead replacing the centrosome (Holy 1997). The model predicts (Fig. 2) that when $L/R = 12$ μm/10 μm = 1.2, the normalized equilibrium distance of the centrosome from the center of the chamber will be $\Delta_e/R \approx 0.4$. In the chamber of the assumed size, therefore, the distance of the centrosome from the center will be $\Delta_e = 0.4R = 4$ μm. This example demonstrates how the unit-invariant form in which the model results are presented can be applied to any specific situation in a quantitative experiment.

It was assumed in the above that all microtubules in the cell have the same length. A generalization of the model to a distribution of lengths is straightforward. To preserve the intrinsic symmetry of the cytoskeleton, the distribution characterized by a density function $q(L)$ should be the same for each orientation of unstressed emanation from the centrosome. The only modification to the model will be to integrate with respect to L in addition to integrating with respect to the emanation angle when finding the total force F. The formula for the total force in the three-dimensional case becomes

$$F = \int f(\theta_0, L)pq(L)\mathrm{d}\Omega\mathrm{d}L$$

Let for example $q(L)$ be the density function of a uniform distribution between $1.05R$ and $1.15R$. Following the same computational strategy, Maly and Maly (2010) find in this case that $\Delta_{eq} = 0.220R$. In the model with the constant length, when its value was equal to the mean of this distribution ($L = 1.1R$), the equilibrium

Fig. 4 Total force exerted by the microtubules on the centrosome for small deviations of the centrosome from the center of a flat cell, starting from the fully symmetric cytoskeleton conformation. Reproduced from Maly and Maly (2010) with permission from Elsevier

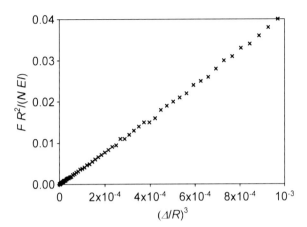

distance was $0.214R$. Thus, the equilibrium distance with the distribution of lengths is not the same but larger, due to nonlinearity of microtubule bending.

The two-dimensional case corresponds to microtubules confined in an essentially flat spatial domain, as in the thinly spread cells cultured on glass in most experiments today. The lowest-energy conformation of the microtubule cytoskeleton, in which each microtubule is in the lower-energy (stable) equilibrium form, can be computed as in the three-dimensional case. It will share the "combed" appearance with the three-dimensional case (see Fig. 3), but this appearance is inconsistent with the images of flat cultured cells, in which neighboring microtubules are typically buckled in opposite directions (Euteneuer and Schliwa 1992). In view of the claim that the centrosome in flat cells is maintained in the geometrical center of the cell outline (Euteneuer and Schliwa 1992; Burakov et al. 2003; Gomes et al. 2005), the conformation of special interest in the flat-cell case is the fully symmetric cytoskeleton. The full symmetry in the appropriate cell-biological sense is a reflection symmetry with respect to any axis that can be drawn through the center of the cell. This implies an infinite-fold rotational, or circular, symmetry, but excludes the case of vortex polarization that the rotational symmetry by itself would permit. The full symmetry thus requires that the centrosome be in the cell center, that the unstrained directions of microtubules be uniformly distributed around the centrosome, and that the two directions of buckling for each unstrained direction be equally represented. It is clear that the fully symmetric conformation is a static equilibrium.

Calculations show (Maly and Maly 2010) that small deviations of the centrosome from the center result in a third-power growth of the total force on the centrosome (Fig. 4). This force is directed outward. Consequently, the symmetry is unstable. The new equilibrium is reached in which the centrosome is removed from the center by a distance approximately equal to the difference of the microtubule length and cell radius (Fig. 2). Precise calculations show that it is slightly smaller than this difference, which means that unlike in the three-dimensional case, all microtubules in the predicted flat equilibrium structure are in contact with the cell boundary and are bent.

As was reviewed in the beginning of this section, the issue of central vs. eccentric positioning of the centrosome has been addressed in studies on flat cultured cells and on biochemically reconstituted microtubule cytoskeletons in artificial chambers. The Maly and Maly model predicts that in a flat arrangement, the deviation of the centrosome from the center under the action of microtubule elasticity alone should be equal approximately to the difference of the microtubule length and the radius of the confining boundary, and that in three dimensions it should be, approximately, twice as large. Direct comparison of this theory with experiments has to date remained impossible. Measurements that would provide the required information are still lacking. Complicating the comparison in the case of the reconstituted systems (Holy et al. 1997; Faivre-Moskalenko and Dogterom 2002) is the small number of microtubules. Due to the smallness of this number, the specific distribution of the few microtubules around the artificial centrosome, and their specific individual lengths affect the positioning greatly. Further, central positioning in the terminology of the reconstitution work at least in some cases described a degree of eccentricity that is significant in the quantitative frame of reference set by the Maly and Maly model. In the absence of measurements, it is unclear whether the terminology of central vs. eccentric positioning is consistent between this quantitative framework and the one that was used qualitatively to classify the positions of centrosomes in the flat cultured cells. The one experimental report (Euteneuer and Schliwa 1992) that specifies the deviation from the center that was still considered as insignificant sets this distance at 5 µm. While the microtubule length and the cell size measurements were not at the same time reported, from the images one might judge the above magnitude of deviation as potentially comparable with the eccentric equilibrium predicted by the Maly and Maly model. The notion that the centrosome is positioned in the geometric center of flat cells, contrary to the quantitative model prediction, should be re-examined by means of measurements done in the quantitative framework set by the model. More generally, the very limited comparison that is currently possible between the quantitative theory and experiment underscores the need for a firm theoretical basis in future work that will address the fundamental problem of the cell body positioning within the cell boundary.

Bipolar Cell Body

The unipolar cell body considered in the last section, which is centered around the single interphase centrosome, becomes bipolar in preparation for division. The division of the cell body is most commonly accompanied or followed by the division of the cell boundary, as defined in the introduction, and together they comprise the division of the cell. The bipolar microtubule cytoskeleton of the dividing cell body (mitotic spindle) in characteristic instances consists essentially of two asters, each assembled around its own centrosome (spindle pole). Positioning of the mitotic spindle through the interaction of the astral microtubules with the cell boundary often determines whether the cell division will be symmetric or asymmetric. This process plays a crucial role in development.

The model considered in the last section was extended (Maly 2012) to calculate the force exerted on the spindle by astral microtubules that are bent by virtue of their confinement within the cell boundary. It was found that depending on parameters, the symmetric position of the spindle can be stable or unstable. Asymmetric stable equilibria also exist, and two or more stable positions can exist with the same parameters. The theory poses new types of questions for experimental research. Regarding the cases of symmetric spindle positioning, it is necessary to ask whether the microtubule parameters are controlled by the cell so that the bending mechanics favors symmetry. If they are not, then it is necessary to ask what forces external to the microtubule cytoskeleton counteract the bending effects sufficiently to actively establish symmetry. Conversely, regarding the cases with asymmetry, it is necessary to investigate whether the cell controls the microtubule parameters so that the bending favors asymmetry apart from any forces that are external to the microtubule cytoskeleton.

Cells often divide symmetrically to produce two daughter cells that are of equal or approximately equal size. Cell lines on which experiments are conducted in cell culture tend to reproduce in this fashion. Asymmetric divisions that produce daughter cells of unequal size abound during development and differentiation in multicellular organisms. Experimentally well-characterized examples include single-cell embryos of the mussel *Unio* (Lillie 1901), roundworm *Caenorhabditis* (Hyman and White 1987) and leech *Helobdella* (Symes and Weisblat 1992), *Drosophila* neuroblasts (McCarthy and Goldstein 2006), and mammalian oocytes (Schuh and Ellenberg 2008). In the roundworm, for example, the first unequal division creates a larger cell that is the first somatic cell and a smaller cell that continues the germ line.

In addition to the significance of size as such, for example between the large stem cell and its small progeny (Watt and Hogan 2000), the division into daughter cells of unequal size may lead to an unequal distribution of specific components of the mother cell cytoplasm between the progeny. Such components may include cell fate determinants (Whittaker 1980). In this connection, the general notion of asymmetric cell division includes cases where the daughter cells are of equal size, yet differ in the complement of components that they inherit. For a broader review of such cases, in addition to the cited work by McCarthy and Goldstein, one may be referred to Knoblich (2008) or Siller and Doe (2009). Here the term "asymmetric division" is used exclusively in reference to the division that generates daughter cells of unequal size. This case presents an obvious challenge for biomechanical explanation.

Generally, cells divide through the middle of the mitotic spindle (Bray 2001). The spindle proper consists of microtubule bundles that connect the two spindle poles. Precise terminology was developed (Maly 2012) in order to formulate the theoretical question posed by the symmetric and asymmetric cell division through the positioning of the bipolar dividing cell body within the still-common cell boundary. The line segment connecting the two spindle poles was termed the physical spindle axis. In the geometrical sense, this axis can be extended to define a coordinate axis that passes through the cell. The instance of the (extended) spindle axis passing through the cell center can be considered first. The paradigmatic cases, e.g., the HeLa cultured cells (Théry et al. 2005) or the first division in *Caenorhabditis* (Grill et al. 2001), seem to exhibit this geometry. The cell center can be defined as the geometrical center of the space delimited by the cell boundary. The fundamental question posed by the symmetric and asymmetric cell division can then be formulated as follows: What determines coincidence of the center of the spindle proper with the cell center? What determines its shift from the cell center along the spindle axis?

Emanating from the spindle-pole centrosomes alongside the microtubules of the spindle proper, astral microtubules radiate to the cell periphery. As reviewed below (see also Pearson and Bloom 2004), it is generally believed that the spindle is positioned through the astral microtubules' interaction with the cell boundary. Three kinds of effects are considered in this connection: the bending elasticity of microtubules that are deformed by their contact with the cell boundary, the stochastic assembly and disassembly of the microtubules, and pulling on the microtubules by molecular motors that are anchored on the cell boundary. It must be observed that the bending elasticity is an intrinsic property—all microtubules possess it. In contrast, the stochastic assembly and molecular motors may or may not play a role, depending on the specific intracellular conditions. Therefore, the latter two mechanisms will always influence the spindle position in combination with the effects of bending elasticity. This natural hierarchy makes it imperative to understand the effects of bending elasticity for the success of integrated biomechanical understanding of symmetry and asymmetry of the bipolar cell body positioning and cell division.

The pioneering theory by Bjerknes (1986) explained the salient features of spindle positioning in amphibian blastomeres. It dealt exclusively with bending elasticity, but microtubule deformations were not computed explicitly. It was assumed that straight astral microtubules radiate in all directions equally from the spindle poles. Each microtubule by assumption contacted the cell boundary with its distal end and developed the Euler buckling force that corresponded to its length. In addition to reproducing the orientations of spindles in the changing cell geometry of the progressively dividing embryo, it was demonstrated that the orientation exhibited bifurcations. Alternative equilibrium orientations appeared with the elongation of the spindle relative to the cell size. The centers of the spindles, however, corresponded with the cell center. An eccentric equilibrium required that the two asters be unequal, for example, in the bending rigidity of their microtubules. The adequacy of the straight-line and buckling-force approximations and the stability of the calculated equilibria to perturbations were not assessed.

Subsequently, in a model by Théry et al. (2005), the orientation of the spindle in a circular cell was treated as resulting from pulling on astral microtubules by postulated force-generating elements anchored on the cell boundary. Only situations with central symmetry were considered. Microtubule deformations and bending elasticity were not included in the model. Force-balance models of pole separation and spindle morphogenesis (Cytrynbaum et al. 2003) treated the astral microtubules similarly. The model by Grill et al. (2001) that addressed the question of the asymmetric first division of *Caenorhabditis* similarly did not include microtubule bending. It was concluded that the shift of the spindle was caused by unequal distribution of the pulling elements on the cell boundary, which therefore exerted unequal force on the two poles. A different model by Grill et al. (2005) dealt with the spindle-pole oscillations, which are observed in the *Caenorhabditis* embryos. The oscillations and their theoretical modeling as such will be reviewed in the last section; here we will be concerned with the general methodology of modeling the deformation of confined microtubules, which was advanced by these models. In the new model by Grill, bending of astral microtubules against the cell boundary was included in addition to the pulling. The microtubule deformations were not explicitly computed. The force associated with them was computed using a linear Hookean dependence on the distance between the pole and the boundary. The resulting oscillations of the pole in this one-dimensional model were about the middle, symmetric position. A different model for the oscillations (Kozlowski et al. 2007) computed the microtubule deformations explicitly in three-dimensions. The deformations were caused by the viscous drag in the cytoplasm and instantaneous pulling by the postulated pulling elements on the ends of the microtubules that were coming in contact with the boundary. Upon contact with the boundary, the microtubules in this model disassembled, preventing development of a durable deformation of bending against the boundary.

It should be noted that in contrast with the reviewed simplifications of the treatment of the astral microtubules in spindle models, the deformations of the microtubules in the spindle proper (those that connect the two poles) have been treated with the precision of the standard bending elasticity theory (Nedelec 2002; Rubinstein et al. 2009).

The model by Maly (2012) extended and complemented these approaches, permitting a general analysis of the effects of the microtubule elasticity on spindle positioning. It focused on the question of symmetric vs. asymmetric positioning of the spindle, as posed above. Deformations of astral microtubules were computed explicitly, and the stability of the equilibria to perturbations was assessed. In view of insufficient experimental data pertaining to quantitative descriptors of mitotic spindles, a theoretical context had to be established in which the experimentally observed structures might subsequently be placed. The model allowed treatment of the different theoretical regimes as well as calculation of some sample structures that was deemed representative.

Instead of one microtubule aster around the single centrosome in the interphase model (Maly and Maly 2010), in the mitotic model (Maly 2012) there are two asters around the two rigidly coupled centrosomes at the spindle poles. These asters may be partial in the sense that the microtubules may not radiate in all directions. Material spindle poles are small compared with the cell size, and the spindle proper which connects them contains a large mass of crosslinked microtubules (Bray 2001). In view of this, in the model the spindle proper is an absolutely rigid segment that connects two points in space that represent the poles (centrosomes). In this respect, the model is similar to the models by Bjerknes (1986), Kozlowski et al. (2007), and Théry et al. (2005). The length of the spindle proper (the interpolar distance) is denoted S.

To address the problem as posed above, axially symmetric situations are considered, in which the axis of symmetry coincides with the spindle axis (i.e., passes through the poles). In the Cartesian coordinate system of the model, the x axis is collinear with the spindle axis. When isolated spindle poles are considered, and unless otherwise noted, the pole on the right is meant. The pole coordinate is denoted x_p; the coordinate of the middle of the spindle proper is x_s. F_p is the projection on the x axis

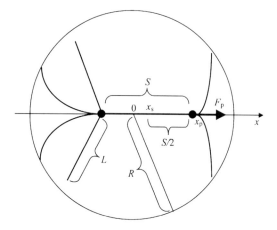

of the total force acting on an isolated pole. F_s is the projection on the x axis of the total force acting on the spindle (i.e., on both poles). Both forces are collinear with the x axis due to the axial symmetry. Figure 5 illustrates the model and notation.

Microphotographs in the papers cited in the introduction confirm that microtubules in mitotic cells do not converge on spindle poles at sharp angles in a fan-like arrangement. This rules out a hinged (free pivoting) boundary condition on the centrosome. The smooth bending forms seen in microphotographs argue in favor of equilibrium flexure and against additional constraints. Accordingly, like the microtubules in the interphase model, the astral microtubules are assumed to be rigidly clamped at the centrosomes, and their contact with the spherical cell surface, frictionless. The lowest-energy equilibrium solution is calculated, because, as discussed in the last section, the alternative equilibria are unstable in three dimensions.

Unlike in the fully isotropic centrosome in the interphase model, the spindle model must take into account the fact that inward-pointing microtubules at the spindle poles may not be astral microtubules but may instead be part of the spindle proper. Accordingly, the angle θ between the clamped direction and the outward direction of the spindle axis in the model does not exceed a certain angle θ_{max}. Whereas in the interphase model N is the number of microtubules in the single microtubule aster, in the mitotic model N denotes the number of microtubules that radiate from one of the two poles. Intrinsically, the model cytoskeleton structure is centrally symmetric: EI, L, N, and θ_{max} are equal on both poles. No assumptions need to be made a priori about parameter values. The model behavior, insofar as it concerns the problem posed here, is controlled entirely by three compound parameters: L/R, S/R, and θ_{max}. Analysis in their space can be for the practical purposes exhaustive.

Limiting cases set the theoretical context for the mechanics of the more biologically relevant regimes. The simplest behavior is exhibited by the structural case of astral microtubules that emanate from the pole exclusively along the axis of the spindle ($\theta_{max} = 0$). As the pole moves collinearly with the spindle axis, the microtubules will abut on the cell boundary, buckle, and bend, exerting force on the pole.

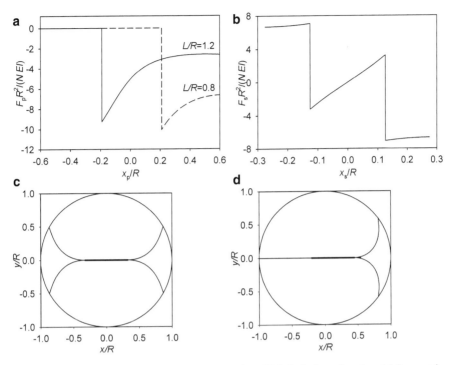

Fig. 6 Limiting case of $\theta_{max} = 0$. (**a**) Pole force function. (**b**) Spindle force function. (**c**) Symmetric equilibrium. (**d**) Asymmetric equilibrium. (**b–d**) $L = 0.8R$, $S = 0.65R$. For clarity, only two microtubule forms are plotted. These microtubules lie in the (x, y) plane that passes through the spindle axis. The circumference is the section of the cell surface, and the thicker line segment depicts the spindle proper. Reproduced from Maly (2012) under the Creative Commons Attribution License

A sample calculation is presented in Fig. 6. When the axial distance of the pole from the boundary is greater than the length of the astral microtubules, the force is zero. When the distance is equal to the length, the magnitude of the force can take any value between zero and the buckling force. For shorter distances, the force decreases, as the increasingly bent microtubules become less efficient at resisting the displacement of the pole. When the movement of two such poles is coupled through the spindle proper, three regimes are possible. If the spindle proper is short, the astral microtubules may not come in contact with the boundary. This happens when $L + S/2 < R$. The symmetric position will then be a neutral equilibrium. The special case of the entire structure just fitting in the cell without deformation is described by $L + S/2 = R$. Naturally, this rather special condition is unlikely to be realized. It can be noted parenthetically that models operating with microtubule buckling forces, such as the pioneering Bjerknes model, presuppose this condition while also implicitly assuming that it can be maintained as the centrosomes shift. This condition is difficult to cast in physical terms even hypothetically. Returning to the model being reviewed here, the non-trivial situation arises when $L + S/2$ is greater than R. In this case, microtubules emanating from one pole or from both must be bent.

Consider the symmetric case, in which both sides are bent equally. This, obviously, is an equilibrium. However, the magnitude of the force exerted on the pole in this situation is a locally decreasing function of the pole's distance from the boundary (Fig. 6a). Therefore movement of both poles to the right will decrease the magnitude of the force exerted by the boundary on the right pole. By symmetry, the magnitude of the force exerted on the left pole will be increased. The symmetric equilibrium proves unstable. The coupled poles will continue moving spontaneously, until the new equilibrium is reached. In this other equilibrium, the force of the bent microtubules acting on one pole is balanced by the force of the straight microtubules acting on the other pole. This is always possible, because the magnitude of the force of a bent microtubule is always lower than the buckling force. The asymmetric equilibrium is stable. Indeed, movement in the direction of the bent microtubules will leave the opposite pole unsupported, as the straight microtubules lose contact with the boundary. Movement in the opposite direction will place the system in a state it already passed during its spontaneous movement from the unstable symmetric equilibrium.

Thus, in this special case (the limiting case of $\theta_{max} = 0$), we observe instability of symmetry, stability of asymmetry, and the possibility to predict the stable conformation from the structural parameters that include the length of the spindle S and the length of the astral microtubules L in relation to the cell radius R. The forces are proportional to the microtubule bending rigidity EI and to the number of the microtubules N, but the positions and stability of the equilibria do not depend on these parameters. The more general and biologically relevant cases exhibit more complex behavior but retain these fundamental characteristics.

The opposite extreme case is also revealing—the special case of complete, intrinsically spherical asters at each pole ($\theta_{max} = \pi$). Now the behavior depends on whether the astral microtubules are longer or shorter than the cell radius. The case of short microtubules is simple. After the aster comes in contact with the boundary, the force exerted on the pole increases gradually with x_p. The graduality is due to the number of the microtubules in contact with the boundary increasing gradually, unlike in the case of $\theta_{max} = 0$. In addition, only the axial microtubules ($\theta = 0$), whose contribution to an aster with $\theta_{max} \neq 0$ is infinitesimal, go through developing the buckling force during the axial movement of the spindle; all others deflect on contact. The bending leads to a decrease in stiffness, as can be seen in Fig. 7. The decrease in the stiffness of the aster (a progressively shallower slope of the force curve) is different from the previous case, where the magnitude of the elastic force was decreasing with the progressing deformation. The numerical results (Fig. 7) indicate that the softening effect of the deformation (Fig. 6) in the case of the complete aster is more than offset by the increasing numbers of microtubules that come in contact with the boundary. Thus, even though the total force is a nonlinear function of the pole position, the force resisting the outward movement of the pole is monotonic. Due to the monotonicity, the stability of symmetry, which would be expected with simple Hookean (linear) elasticity, is exhibited by a spindle with two complete asters of short microtubules. In addition, the monotonicity means that there is only one equilibrium conformation of the mitotic microtubule cytoskeleton, insofar as the latter is large enough to maintain contact with the cell boundary.

Fig. 7 Limiting case of $\theta_{max} = \pi$, short microtubules. (**a**) Pole force function. (**b**) Spindle force function. (**c**) Equilibrium conformation. $L = 0.8R$, $S = 0.65R$. For clarity, only few microtubule forms are plotted. These microtubules lie in the (x, y) plane that passes through the spindle axis. Their values of θ are sampled uniformly between 0 and θ_{max}. Reproduced from Maly (2012) under the Creative Commons Attribution License

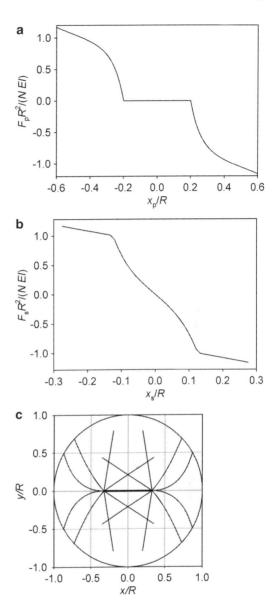

The positioning of an isolated complete aster with comparatively long microtubules in three dimensions was considered in the previous section (Figs. 2 and 3). In this case, considered now as a model for an isolated spindle pole, the pole force function has a root at approximately $x_p = 2(L - R)$: A more centrally positioned pole is attracted to the cell margin, whereas a more eccentrically positioned one is repelled. Considering a spindle with such an aster at each pole, one can observe that the presence of the root does not affect the stability of symmetry. Even though the pole force

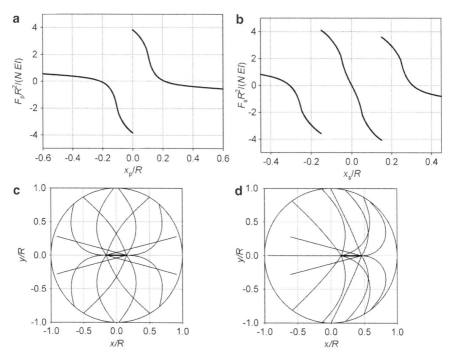

Fig. 8 Limiting case of $\theta_{max} = \pi$, long microtubules. (**a**) Pole force function. (**b**) Spindle force function. (**c**, **d**) Equilibrium conformations. $L = 1.1R$, $S = 0.3R$. Plotting conventions as in Fig. 7. Reproduced from Maly (2012) under the Creative Commons Attribution License

function changes sign (Fig. 8a), it is monotonically decreasing for $x_p > 0$ like it did in the case of short microtubules. Thus, the symmetry of the spindle will be stable.

Overall, however, the pole force function with long microtubules is not monotonic (Fig. 8a). This opens the possibility for existence of additional equilibria, when both poles are on the same side of the cell. The system remains attracted to the symmetric equilibrium until the spindle proper is moved entirely into one half of the cell. Then the conformational transition that was described in the last section in connection with the single three-dimensional aster occurs in the more central aster of the spindle. It places the mitotic system on the other branch of the spindle force function (Fig. 8b). The monotonicity of each half of the single pole force function ensures that following this transition, the spindle continues moving spontaneously in the direction that was previously forced. Indeed, even if the more distal pole is already in the region where it is repelled by the boundary, the more central pole upon crossing the center always experiences a greater force repelling it from the center. The new equilibrium will be reached when the two forces become equal in magnitude. This strongly asymmetric equilibrium will be stable. This is due to the piecewise monotonicity: Whether the poles are on different sides or on the same side of the cell, the force function (Fig. 8a) is locally decreasing for each pole.

Fig. 9 Transition between $\theta_{max}=0$ and $\theta_{max}=\pi$ in the case of short astral microtubules. $L=0.8R$. (**a**) Pole force function. (**b**) Spindle force function, $\theta_{max}=\pi/5$. Reproduced from Maly (2012) under the Creative Commons Attribution License

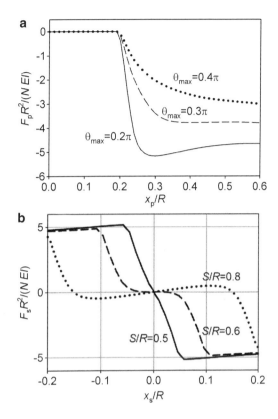

A centripetal movement of the spindle from the asymmetric equilibrium increases its repulsion from the center and decreases its repulsion from the nearest boundary, and vice versa. The spindle will therefore return to the asymmetric equilibrium, unless the forced movement brings one of the poles to the opposite half of the cell, in which case it will spontaneously continue toward the symmetric equilibrium.

Transitions between the limiting cases further illuminate the systems-biomechanical underpinnings of the spindle positioning. Let us first consider the instance of short astral microtubules. We have seen that as θ_{max} increases from 0 to π, stability of symmetry and nonexistence of asymmetric equilibria replace instability of symmetry and stability of asymmetry. Computations demonstrate that increasing θ_{max} from zero first makes the development of the extremum pole force smooth (Fig. 9a, solid curve; cf. Fig. 6). The finite interval of x_p in which the pole force function is decreasing emerges immediately when θ_{max} exceeds zero. The range of values of S that place symmetric poles in these intervals will correspond to stable symmetry. Unlike in the extreme case of $\theta_{max}=\pi$, however, the pole force function is increasing beyond the extremum (Fig. 9a). Let us denote the pole position that corresponds to the extremum x_p'. For S exceeding $2x_p'$, the behavior seen with $\theta_{max}=0$ is retained, and symmetry is unstable (Fig. 9b). The corresponding stable asymmetric equilibrium is derived from the one described in the case of $\theta_{max}=0$: One pole

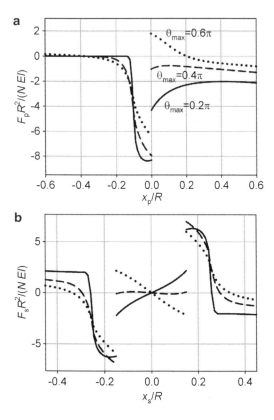

Fig. 10 Transition between $\theta_{max}=0$ and $\theta_{max}=\pi$ in the case of long astral microtubules. $L=1.1R$. (**a**) Pole force function. (**b**) Spindle force function. $S=0.3R$. The line styles correspond to values of θ_{max} as in (**a**). Reproduced from Maly (2012) under the Creative Commons Attribution License

occupies a position that lies toward the cell center from the pole force function extremum, and the other occupies a position toward the cell margin from that pole's extremum. With a further growth of θ_{max}, the aster increases resistance at more eccentric pole positions (Fig. 9a, dashed curve). This eventually erases the non-monotonicity of the pole force function. Figure 9a shows that for sufficiently large θ_{max} the function becomes monotonic as in the case of $\theta_{max}=\pi$ (cf. Fig. 7). The spindle now possesses only one equilibrium, which is symmetric and stable. Thus with short astral microtubules, there is an abrupt transition at some intermediate θ_{max} to the behavior that is qualitatively like the one observed with $\theta_{max}=\pi$.

With long microtubules, the increase of θ_{max} at first leads to the same changes as with the short microtubules, namely the smoothing of the pole force function extremum (Fig. 10a, solid curve; cf. Fig. 6). In addition, the function in this case develops a discontinuity. It is caused by the conformational transition between the lowest-energy conformations that differ on the two sides of $x_p=0$, as discussed in the previous section in connection with the interphase model. The transition was non-observable in the case of $\theta_{max}=0$ because the alternative conformations in that degenerate case were equivalent. A more consequential difference from the case of the short microtubules is that the long microtubules are more deformable, and the drop of the force magnitude past the buckling-force maximum that they exhibit is deeper. Because of this, the increase in resistance at large pole displacements in this case cannot erase the nonmonotonicity efficiently. Instead, at around $\theta_{max} = \pi/2$ the pole force function develops two descending branches (Fig. 10a). The difference from the already considered limiting case of complete asters (Fig. 8a) is that the pole force function with the intermediate θ_{max} may not change sign across the discontinuity. This reflects the comparatively simple behavior of asters with $\theta_{max} < \pi/2$: The individual pole would find equilibrium when all microtubules that are emanating from it are straight, unlike in the more complete asters, where bending is unavoidable. Despite this difference in the individual behavior of separate poles, poles coupled by the spindle will behave in the same way as with the more complete asters of long microtubules, because of the fundamental similarity of the pole force function with the two descending branches. The symmetric position is stable, as is the asymmetric equilibrium (Fig. 10b, dashed curve).

The intermediate case $L \approx R$ presents special interest. Theoretically, it connects the qualitatively different types of behavior described for long and short microtubules. Biologically, astral microtubules seem to be comparable in length with the cell radius in the morphogenetically important instances of large cells of early embryos (Lillie 1901; Hyman and White 1987; Symes and Weisblat 1992; Grill et al. 2001). Evidently, the case of $L = R$ is by itself unrealistic, because the two quantities cannot be exactly equal. This special case, however, establishes a useful reference in the space of the model parameters and regimes of behavior.

Calculations show that a complete ($\theta_{max} = \pi$) aster with $L = R$ develops the peak force just before its pole reaches the cell center (Fig. 11a). This force is associated with the buckling force of the microtubules that all straighten when the pole reaches the center. When the pole is in the center, the elastic force can take any value between the positive and negative extremum, and will be zero for an aster not subjected to any external force. Thus, an individual separate aster exhibits a special kind of stability of the central position, in which a finite restoring force develops upon an infinitesimal perturbation.

However, the magnitude of the restoring force decreases with the magnitude of the perturbation (Fig. 11a). A pole that is closer to the center therefore experiences a higher centripetal force than a pole farther away from the center. This is a condition for instability of symmetry of the spindle. One pole's centripetal movement will be completed at the expense of the increasingly eccentric position of the coupled pole. This is demonstrated by the spindle force function (Fig. 11b). In the asymmetric equilibrium, which is stable, the force on the eccentric pole is balanced by an equal force supported by the straight microtubules of the centrally positioned pole.

The described behavior is observed for spindles with interpolar distances not exceeding a certain value. Inspection of the pole force function (Fig. 11a) shows that the stiffening of the aster at large eccentricities of the pole creates a range of extreme pole positions where the magnitude of the force is increasing with the distance from the center. This creates conditions for stability of symmetry of sufficiently long spindles (Fig. 11c).

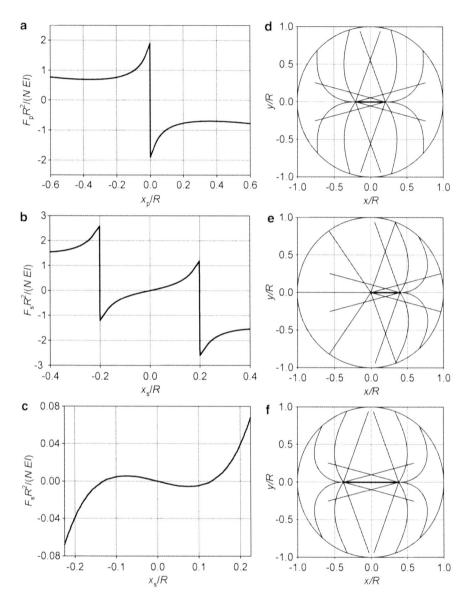

Fig. 11 Special case of $L=R$, $\theta_{max}=\pi$. (**a**) Pole force function. (**b**) Spindle force function, $S=0.4R$. (**c**) Spindle force function, $S=0.75R$. (**d**) Unstable symmetric equilibrium, $S=0.4R$. (**e**) Stable asymmetric equilibrium, $S=0.4R$. (**f**) Stable symmetric equilibrium, $S=0.75R$. Reproduced from Maly (2012) under the Creative Commons Attribution License

The domain of instability of the spindle symmetry in the three-dimensional space of the structural parameters (θ_{max}, S/R, L/R) is outlined in Fig. 12. The behavior described in the preceding sections can be seen in its outline alongside some additional features. The boundaries of the domain show how the continuity between the regimes characteristic of the short and long microtubules is achieved through the

Fig. 12 Symmetry instability domain. In the three-dimensional space of structural parameters (θ_{max}, S/R, L/R), L/R is understood as the vertical dimension in this graph. Isolines corresponding to the indicated values of L/R are shown. (**a**) Surface bounding the domain from below. (**b**) Surface bounding the domain from above. Reproduced from Maly (2012) under the Creative Commons Attribution License

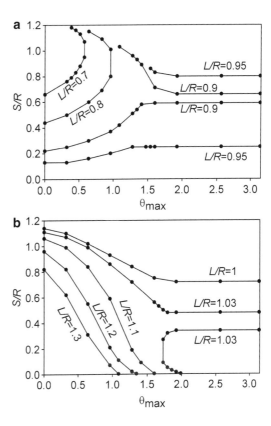

special case of $L=R$. Specifically, for both long and short microtubules, the (θ_{max}, S/R) cross-sections of the symmetry instability domain are restricted to $\theta_{max} < \pi/2$.

For short microtubules, longer spindles exhibit unstable symmetry. Note, however, that the stability of spindles that are particularly short ($S < 2(R-L)$) is only neutral in this case, because in their symmetric position, the astral microtubules do not contact the boundary. For long microtubules, instability of symmetry is exhibited by shorter spindles. In this case, it is of note that the longer spindles exhibit bistability between the symmetric and asymmetric equilibria, as was illustrated in Fig. 10b (dashed curve).

For $L \approx R$, the range of instability of symmetry extends through $\theta_{max} = \pi$. This behavior, which has been illustrated in the special case of $L=R$, is retained for L taking values between approximately 0.9 and 1.05R. Although this range is narrow in absolute terms, it seems to be particularly relevant, because inspection of images in the literature suggests that L is often similar to R in the morphogenetically important instances of large cells in early embryos (Lillie 1901; Hyman and White 1987; Symes and Weisblat 1992; Grill et al. 2001).

It appears plausible that if each microtubule length from a set supports stability of symmetry, then any distribution of lengths over that set will also support stability of symmetry. Conversely it may be posited that if each length by itself supports instability, any distribution will also support instability. Obviously, if any subset supports neutral stability, the outcome will be dictated by the complementary sub-set. With this in mind, predictions about the stability of symmetry with microtubule length distributions can be obtained by mere inspection of the stability and instability domains already mapped.

The instability domain in the (θ_{max}, S/R, L/R) space is bounded and embedded in the domain of stability (Fig. 12). Furthermore, it lies above the domain of neutral stability ($S/R < 2-2L/R$). At large θ_{max}, there is a considerable separation between the domains of instability and neutral stability; at small θ_{max}, they almost touch.

The support of a microtubule length distribution is represented in this parameter space by a vertical segment. If such a segment lies entirely within or outside the instability domain, the central symmetry will be, correspondingly, unstable or sta-ble. For example, any distributions whose support falls entirely to the right of the $L/R = 1$ isoline in Fig. 12b will predict stable central positioning of the spindle.

Further, unless the distribution is unusually sharply concentrated at the interme-diate values, the instability domain can be considered as touching the neutral stabil-ity domain for smaller θ_{max}. Therefore, distributions whose support falls below the upper bounding surface of the instability domain (Fig. 12b) predict instability of symmetry, if θ_{max} is not large.

In the likely case of descending exponential distributions (Gliksman et al. 1992), the contribution of the few exceptionally long microtubules may prove negligible. In that case, the prediction can be further simplified: For larger θ_{max} and S/R, stabil-ity is predicted, and for smaller θ_{max} and S/R, instability is predicted.

Among the theoretically possible structures and equilibria, several can be con-sidered paradigmatic, based on qualitative examination of images of spindles in the experimental literature. Firstly, there is the case of a long spindle with short astral microtubules that radiate from the poles in a wide angle. The equilibrium conforma-tion is plotted in Fig. 13a. According to the preceding analysis, in this regime (large S/R, small L/R, large θ_{max}), the symmetric equilibrium is the only equilibrium, and it is stable. Awaiting measurements motivated by the theory, it can be said that this regime appears common among the variety of equally dividing cells. The HeLa cultured cells are one example (Théry et al. 2005).

Another characteristic example is the structure with long astral microtubules that radiate in a wide angle. This regime exhibits the bistability between the symmetric and asymmetric equilibria. The alternative conformations are illustrated in Fig. 13b, c. This example is characterized by a comparatively small S/R and large L/R. In this respect it is reminiscent (in the asymmetric conformation) of the first division in the invertebrate models of development that include the mussel *Unio* and the roundworm *Caenorhabditis* (Lillie 1901; Hyman and White 1987; Symes and Weisblat 1992; Grill et al. 2001). For comparison, Fig. 13d displays the only stable equilibrium that exists in the regime with $L \approx R$ and a spindle of medium length, which is asymmetric. In terms of the spindle proportions and position, this last example is reminiscent of the mouse oocyte (Schuh and Ellenberg 2008).

The example of bistability in particular raises the question of the absolute mag-nitude of the collective spindle forces. The natural unit of force, NEI/R^2, equals 22 pN when $N = 100$, $EI = 26$ pN μm^2, and $R = 10$ μm. The barrier for switching from

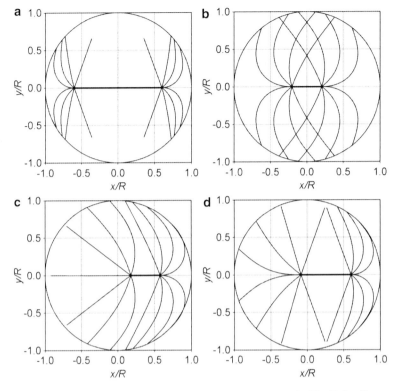

Fig. 13 Sample equilibrium conformations. Plotting conventions as in Fig. 7. In all examples, $\theta_{max} = 0.6\pi$. (**a**) $L=0.7R$, $S=1.2R$. (**b, c**) represent the alternative conformations that exist with $L=1.1R$ and $S=0.4R$. (**d**) $L=0.95R$, $S=0.7R$. Reproduced from Maly (2012) under the Creative Commons Attribution License

the symmetric to asymmetric state in the case plotted in Fig. 10b (dotted curve) is then 65 pN. Note that the barrier for switching from the asymmetric state is higher, 156 pN. The above estimates were order-of-magnitude for N and R in a generic cell, and a mid-range experimental value was used for EI (Mickey and Howard 1995). Equally relevant may be the values $N=1,000$ and $R=5$ μm, in which case the two barriers will be 2.6 and 6.2 nN. Targeted measurements are needed to determine if such barriers (~0.1–1 nN) can be overcome in a given cell type.

So far we have considered the position of the spindle proper (the segment connecting the spindle poles) along the axis of symmetry of the spindle, x, characterizing it by x_s, the coordinate of the center of the spindle proper. The analysis of forces during displacement of the spindle proper along x demonstrates that the central symmetry can be stable or unstable with respect to axisymmetric perturbations. It can also be demonstrated that the system displays the same type of behavior with respect to nonaxisymmetric perturbations, specifically that the axial symmetry can be stable or unstable (depending on the structural parameters) with respect to

Fig. 14 Spindle force
as a function of displacement
orthogonal to the spindle.
(a) $L=0.7R$, $S=0.8R$, $\theta_{max}=\pi$.
(b) $L=1.1R$, $S=0.3R$, $\theta_{max}=\pi$

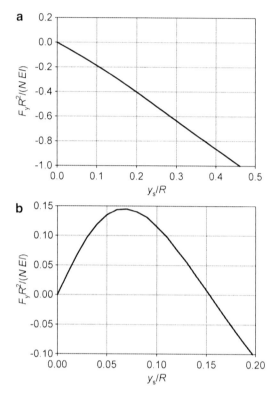

displacement along the orthogonal coordinate axis y. The relevant coordinate of the center of the spindle proper can be denoted y_s, and the projection of the total force acting on the spindle proper, F_y.

Figure 14a demonstrates that long spindles with short astral microtubules exhibit stability of the axial symmetry. Figure 14b demonstrates that short spindles with long astral microtubules exhibit instability of the axial symmetry and stability of a nonaxisymmetric equilibrium. Note that in the limit of short spindles, the model becomes identical with the single interphase aster, and also identical with the axisymmetric spindle model taken to the same limit.

The results pertaining to nonaxisymmetric conformations lend support to the general form of the conclusions that can be drawn about symmetry and stability of static equilibria of spindles and bipolar cell bodies confined by the cell boundary. Collectively the computational results call attention to the theoretical fact that intrinsically symmetric assembly of the spindle microtubules in the absence of external forces will not necessarily result in symmetric positioning of the spindle within the cell boundary. This general conclusion does not apply exclusively to the cases of symmetric vs. asymmetric positioning of the spindle midplane, with which the above exposition has been mainly concerned due to its importance for the fundamental problem of equal vs. unequal cell division, i.e., of symmetric vs. asymmetric position of the dividing, bipolar cell body within the still undivided cell boundary along their common axis of symmetry.

The developed theory of the spindle (Maly 2012, and the nonaxisymmetric calculations presented here) derives the static equilibria from the considerations of bending of the astral microtubules against the cell boundary, and assesses the stability of the equilibria. In this respect, it is an extension of the pioneering spindle model by Bjerknes (1986). The distinctive new method is explicit computation of the bent microtubule forms and stability analysis, like in the application to the interphase cytoskeleton (last section). The chief general prediction obtained with this method is that an intrinsically symmetric mitotic microtubule cytoskeleton may spontaneously adopt asymmetric conformations, when constrained within the cell. In this respect, the model is conceptually derivative from the treatment of the interphase microtubule cytoskeleton that was developed in the last section and from the pioneering work of Holy (1997) on the interphase microtubule asters. The mitotic model reveals the differences of the mechanics of two coupled asters of microtubules that are found at the two poles of the mitotic spindle. The individual confined asters in the interphase models always break the symmetry with respect to the cell center. The mitotic model, at least under certain conditions, exhibits stable equilibria that are centrally symmetric, as well as bistability between the symmetric and asymmetric equilibria.

The model does not distinguish between equilibrium structures that are identical but for rotation with respect to the cell center, because microtubules in such structures are bent equally against the spherical cell outline. When the cell shape deviates from the sphere, the model should apply qualitatively insofar as the position of the spindle along the common axis of the spindle and the cell outline is concerned. An example is the ellipsoidal egg of *Caenorhabditis*. The model is more quantitatively applicable to the other mentioned cases of the first divisions in eggs that are approximately spherical.

The theoretical possibility of the spontaneous development of asymmetry through bending of astral microtubules, and the existence of special requirements for the stability of symmetric conformations, pose new types of questions that can be asked when designing and interpreting experiments. Symmetric spindle positioning cannot be considered a "default" state of the system. Just as importantly, the source of the asymmetry should not be sought necessarily outside an intrinsically symmetric structure consisting of the mitotic microtubule cytoskeleton and the confining cell boundary.

When studying a case of symmetric positioning, it is worth investigating what makes it symmetric. Do the parameters such as the length of the astral microtubules (L) and of the spindle proper (S) have values that support the stable symmetry? If they do not, what forces external to the microtubule cytoskeleton act against the collective bending forces and actively establish the symmetry?

When studying a case of asymmetric spindle positioning, it is worth investigating the possible contribution of the collective bending effects of the astral microtubules to the generation of asymmetry. In fact, in the light of the theory this question acquires priority. Although the quantitative nature of the collective bending that breaks symmetry is complex, the hypothesis that the source of asymmetry resides in the basic cytoskeleton structure itself is simple, compared with hypotheses that involve asymmetric regulation or asymmetric external forces applied to the structure.

The spontaneous development of asymmetry through unequal bending does not by itself have a preferred direction: Each asymmetric equilibrium in the model has a counterpart, which is mirror-symmetric about the cell center. External forces and regulation mechanisms may be responsible for biasing the spontaneous symmetry-breaking, even if they are not responsible for the generation of asymmetry. Similarly, the action of the external forces, even transient, may be responsible for selecting between the symmetric and asymmetric equilibria in the cases of multistability.

Summarizing, the explicit numerical treatment of bending in systems of linked and confined microtubules that was developed by Maly and Maly indicates existence of new types of collective mechanical behavior in confined microtubule cytoskeletons, which include symmetry-breaking and multistability. New types of questions can be asked in experimental work in the light of the theory, which also establishes a quantitative framework that can guide experiment design. Interpretation of new experimental results, and, possibly, re-interpretation of those previously obtained, will require generalization of the numerical model and its rigorous adaptation to the structural features of each cell type.

Boundary Dynamics

In the models presented so far the cell boundary was considered as a constraint on the dynamics of the cell body. The simplification of unchanging boundary revealed the complexity of the system-level mechanical behavior of the constrained microtubule cytoskeleton and uncovered the autonomous capacity of the mechanics of the confined cell body for generation of asymmetric cell conformations. Especially in single, isolated animal cells, however, the boundary is far from rigid. The question that can be asked in the light of the body mechanics models presented in the previous sections is as follows: Does a dynamic boundary participate in the symmetry breaking, or does it counteract this tendency of the confined cell body? This question can be addressed by examining the combination of the developed physical models for the cell body and boundary.

In an ingenious experiment, microtubule asters assembled on artificial centrosomes were placed inside lipid bilayer vesicles (Pinot et al. 2009). The numerical models developed in conjunction with these experiments, however, included only rigid boundaries, like in the other models of confined microtubule cytoskeletons that were reviewed in the previous sections. Numerical treatment of nonconstant boundaries constraining microtubules requires a special approach. When constructing such a model for the cell rather than the simplified experimental model, one has to include the mechanical properties of the boundary that is more complex than the bilayer membrane shell.

The cell boundary mechanics was captured in an abstracted two-dimensional form in models that aimed to predict self-organization of multiple adjacent cell boundaries into nonrandom cellular arrangements characteristic of many tissues (Käfer et al. 2007; Farhadifar et al. 2007). The deformation mechanics of the single animal cells can be described as governed by line tension (Evans and Yeung 1989). This force was found insufficient in the cited computational models for tissue organization, and realistic predictions were obtained by adding phenomenological elastic terms to the effective energy function alongside the line-tension terms. These postulatory terms are not needed in a theory that includes a mechanistic model of the elastic cytoskeleton of the enclosed cell body. Also, to complete the physical description of the boundary dynamics per se, one must include the quasi-elastic effects of oncotic pressure which is associated with the volume enclosed by the boundary.

A theoretical model with the requisite characteristics was constructed by Arkhipov and Maly (2006a) and analyzed further by Baratt et al. (2008). The model minimizes an energy function of a free-floating cell E_f comprised of the microtubule bending energy E_b, surface energy of the boundary E_s, and oncotic volume energy E_v:

$$E_f = E_b + E_s + E_v$$

E_b is obtained by integration of squared curvature along each microtubule and summation over the microtubules in the cell, with the microtubule bending rigidity as a factor. E_s is calculated as a product of the measured cortical tension of unattached cells and the area of the boundary enclosing the microtubules. E_v is calculated using the measured oncotic pressure. The extracellular pressure is assumed to remain constant as the cell volume changes. The work against it that is associated with the changing volume enclosed by the boundary is trivial to calculate. The intracellular oncotic pressure changes with the changing concentration of macromolecules, to which the cell boundary is impermeable. It therefore changes inversely proportionally to the volume. The term of the volume energy function that is the work of the intracellular oncotic pressure when the cell volume V deviates from the oncotic equilibrium volume V_{eq} is then $\Pi V_{eq} \ln(V/V_{eq})$, where Π is the equilibrium oncotic pressure. The calculations of Arkhipov and Maly and Baratt et al. assumed that the extracellular pressure equals the equilibrium oncotic pressure, as is probably the case in mammalian tissues and in blood or lymph. In the future the model could be adapted to specific extracellular pressures to capture, for example, the conditions of experiments in cell culture more precisely.

The model is spatially discretized and the conformation corresponding to a minimum of E_f is found numerically. Multiple runs of the minimization algorithm starting from a random set of conformations converge, depending on the physical parameters, around one conformation or several very similar conformations. With the free

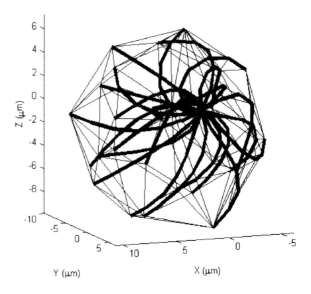

Fig. 15 Cell conformation that minimizes the combined energy of the body cytoskeleton and the boundary. *Bold lines* represent microtubules. The cell boundary is triangulated. Reproduced from Arkhipov and Maly (2006a, doi:10. 1088/1478- 3975/3/3/006) with permission from IOP Publishing

parameters describing the individual cell (the number of microtubules and the oncotic equilibrium volume) taking realistic values, the conformation minimizing the total energy of the cell body and boundary is asymmetric: The centrosome is off center within the boundary and the microtubules are bent unequally (Fig. 15).

This cell boundary model, which takes into account not only the surface area and tension but also the enclosed volume and oncotic pressure, can also be used to answer the question of how the intrinsically undefined orientation of the spontaneously established asymmetry can be influenced by the cell's environment. This question arises, for example, in the biology of lymphocytes, whose intrinsically asymmetric cell body acquires a nonrandom orientation on contact with an infected cell, directing the secretion of the immune response mediators at this cell from the lymphocyte's centrosome region. To assess the relative energy of the cell boundary corresponding to the different orientation of the asymmetric cell body, the approach taken in the last two sections can be reversed: The shape of the cell body can be modeled as a constraint for the dynamics of the cell boundary.

Using this approach, Arkhipov and Maly (2006a) and Baratt et al. (2008) calculated the energy landscape of all possible orientations of the body in a cell whose boundary is deformed by the contact with the target surface. The lowest point on this landscape corresponds to the preferred orientation, which under realistic assumptions about parameters turns out to be one with the centrosome facing the contact zone (Fig. 16). Moreover, multiple minima on the landscape correspond to subpopulations of cells that are numerically predicted to coexist under the same conditions (Fig. 17). These subpopulations, which differ by the centrosome orientation, find their counterparts among cell populations that prove heterogeneous in experiments (Baratt et al. 2008). Future analysis of the systems biomechanics of the

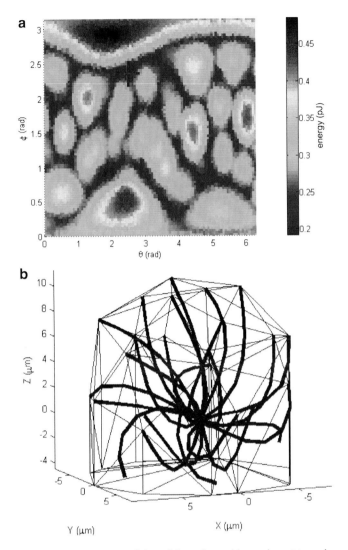

Fig. 16 Minimization of the energy of the cell boundary with attachment to a planar substrate. (**a**) The energy landscape. The angles on the axes define the orientation of the cell body. The global minimum is marked. (**b**) The conformation corresponding to the global minimum. Plotting conventions as in Fig. 15. Reproduced from Arkhipov and Maly (2006a, doi:10.1088/1478-3975/3/3/006) with permission from IOP Publishing

cell will likely explore the dynamic interactions of the cell body and boundary. The simplified approach of considering one as a constraint for the other, however, may still remain valuable conceptually even after the methods of numerical treatment of the fully dynamic interaction become improved.

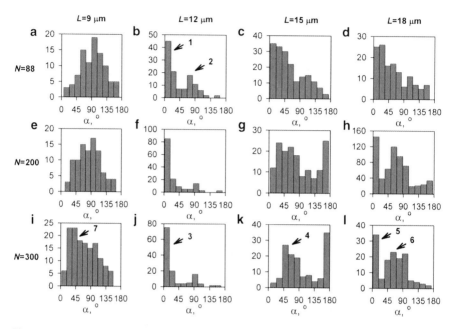

Fig. 17 Orientation of the centrosome with respect to the attachment substrate as predicted by the cell boundary energy model. 0 is orientation with the centrosome toward the substrate, 180° is away from the substrate. Parameter values (L and N) mark the rows and columns of graphs. Experimentally relevant subpopulations are indicated by numbers. Reproduced from Baratt et al. (2008) under the Creative Commons Attribution License

Emergent Irreversibility

Irreversibility is the hallmark of life that has long been recognized from the perspective of biochemical thermodynamics. That energy derived from the nonequilibrium chemistry can be dissipated in the mechanical manifestations of life, such as movement against viscosity and friction is obvious. This section considers some examples that demonstrate that the cell mechanics exhibits also nontrivial system-level irreversibility that emerges specifically on the cellular level of structural organization. In the light of these examples, the cell body–boundary mechanical interaction appears in the general case irreversible. It may be posited that the irreversibility of this type is responsible for the characteristic unidirectionality of the development of biological form.

Let us recall from the previous section that in an effectively flat cell, displacement of the centrosome from the center leads to deformation of microtubules that adopt stable and metastable equilibrium forms. Examples of such forms are shown in Fig. 18. Calculations (Maly and Maly 2010) show that with an increasing deviation of the centrosome from the center, a position will be reached beyond which the metastable equilibrium form does not exist anymore. When this happens, the

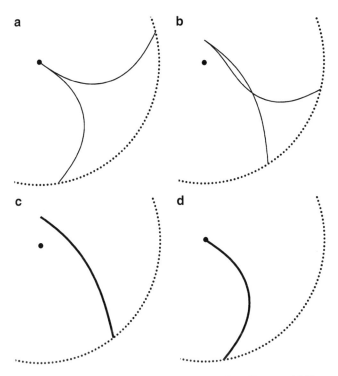

Fig. 18 Equilibrium forms of a microtubule. The *dot* marks the cell center. (**a**) The centrosome is in the cell center. (**b**) The centrosome has been moved away from the center; the two alternative forms are no longer equivalent. (**c**) A further displacement of the centrosome has led to disappearance of the metastable form. (**d**) The centrosome has been moved back to the cell center from the position shown in (**c**); only one of the two potentially existing forms is occupied (cf. **a**). Reproduced from Maly and Maly (2010) with permission from Elsevier

microtubules will adopt the corresponding stable form. "Flipping" into the only remaining equilibrium is an energy-dissipating process. Therefore it cannot be reversed by returning the system to the domain of existence of the metastable solution. This phenomenon underlies our first example of emergent irreversibility in an intrinsically elastic cell body, which is caused by its confinement within the cell boundary.

Irreversibility accompanies the spontaneous evolution of the symmetric flat microtubule aster in response to a perturbation of the centrosome position. The third-power growth of the centrosome-displacing force that the microtubule system develops for small deviations of the centrosome from the center was considered in the previous section (Fig. 4). The system's evolution is reversible for such small deformations of the fully symmetric structure. However, the new equilibrium is reached only when the centrosome becomes removed from the center by a distance approximately equal to the difference of the microtubule length and cell radius (Figs. 2, 19b, and 20). Metastable forms of microtubules gradually become lost beginning with the first inflection point on the forward branch of the force function (solid curve in

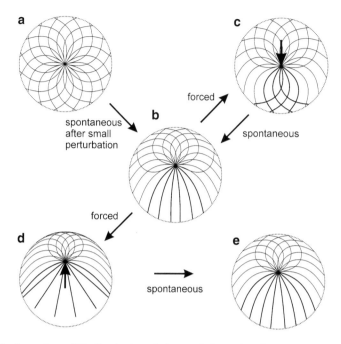

Fig. 19 Conformations of idealized microtubule cytoskeletons in a flat cell. (**a**) Fully symmetric cytoskeleton. (**b**) Equilibrium reached following a small perturbation. (**c**) Central centrosome position restored by an external force. (**d**) Additional deformation caused by an external force. (**e**) Equilibrium reached from the configuration in (**d**) after removal of the external force. Double-width lines show forms that are doubly represented. Reproduced from Maly and Maly (2010) with permission from Elsevier

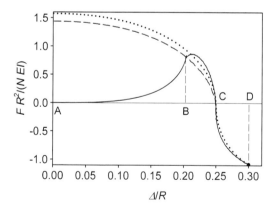

Fig. 20 Total force exerted by the microtubules on the centrosome as a function of the centrosome position in a flat cell. The variables are as defined in Table 1. The *solid curve* is traced in the course of a continuous displacement of the centrosome away from the cell center, starting in the fully symmetric state. The *dashed* and *dotted* curves are traced during forced reversed displacement. Reproduced from Maly and Maly (2010) with permission from Elsevier

Fig. 20, point B). Beyond this point, a reversal of the outward movement of the centrosome places the system on a different branch of the force function. There is a separate reverse branch for each reversal point in the continuous model of Maly and Maly (2010), which makes the force function infinite-valued. A discrete model would have the number of reverse branches on the order of the number of microtubules, which is hundreds in a typical flat cultured cell. In general the force, according to the Maly and Maly model, is a functional of the centrosome path.

The nondimensional form adopted in Fig. 20 reveals the parameter-independent behavior of the model; refer to Table 1 in the last section for nomenclature. The plotted quantity $FR^2/(N\text{EI})$ is the average force per microtubule (among the N microtubules in the cell), expressed in the natural units of force in the model. The natural unit of force is the flexural rigidity of a microtubule EI divided by the square of the cell radius R. Assuming, for example, a rigidity of 25 pN μm^2, which would be near the midpoint of the range of the measured values (see a compendium table in Kikumoto et al. 2006), the force unit will be 1 pN for a cell that is 10 μm in diameter. In this way, the force acting on the centrosome can be estimated in common units when the parameters (N, R, and EI) characterizing the individual cell are known.

Continuing the concrete experimental example from the previous section, assume now that the containing artificial chamber was prepared in such a way that it is shallow (Holy et al. 1997), and the microtubules are 12.5 μm long. In this case, using Fig. 2 and the calculation strategy presented in the last section, we derive that the equilibrium distance of the artificial centrosome from the center is predicted to be very close to 2.5 μm. If an optical trap is then used to displace the bead serving as the artificial centrosome from this position to 2 μm from the center, then the model predicts that the force exerted by the trap on the bead will be very close to 1 $N\text{EI}/R^2$, as can be read directly from the plot in Fig. 20. To compare this prediction with the force as measured in common units by the optical trap technique, one should substitute the values of N, EI, and R that characterize the specific experiment. Using the above experimental estimate for the microtubule rigidity, we obtain 1 $N\text{EI}/R^2 = 20 \times 25$ pN $\mu m^2/(10~\mu m)^2 = 5$ pN.

As examples in Fig. 20 illustrate, the reverse branches are non-zero at the zero centrosome displacement. Thus, although the central position of the centrosome can be restored by forces external to the microtubule cytoskeleton, the new central position will not be an equilibrium (Fig. 19c). The symmetry loss in response to a small perturbation proceeds spontaneously beyond the range of reversibility. One may also ask what happens if the centrosome is forced beyond the spontaneously achieved equilibrium (Fig. 19d). Calculations show that the reverse branches from beyond the equilibrium point pass very close, within 1 %, of the original equilibrium (see the example in Fig. 20). Therefore, the irreversible effect of the perturbation will not be detected by recording the equilibrium centrosome position, and the eccentric equilibrium position of the centrosome in flat cells may, for practical purposes, be called stable. Even through the centrosome position is almost precisely restored after removal of the external force, forcing the centrosome farther away from the center leaves an irreversible trace in the cytoskeleton structure (Fig. 19e).

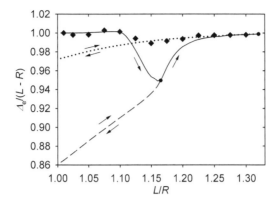

Fig. 21 Equilibrium position of the centrosome as a function of microtubule length in a flat cell. *Diamonds* show positions reached after a small perturbation of a fully symmetric cytoskeleton. *Lines* show changes in the course of continuous elongation and shortening of microtubules. Reproduced from Maly and Maly (2010) with permission from Elsevier

Taken together with the irreversible loss of symmetry in response to infinitesimal perturbations, this permits a generalization: The equilibrium form of a microtubule cytoskeleton confined in a flat cell exhibits memory of past perturbations.

The diamonds in Fig. 21 plot the positions to which the centrosome moves spontaneously in response to small perturbations of the fully symmetric cytoskeleton with the given ratio of the microtubule length to the cell size. The plot demonstrates that although the relationship $\Delta_e = L - R$ is not exact, it is a useful approximation. The sample points comprising this plot are not connected by a curve. This feature of Fig. 21 is meant to reflect the fact that each equilibrium is a result of a separate process of relaxation from the fully symmetric state. Because of the already noted strong nonlinearity, the system will not move from one of these points to the other, when the microtubule length is varied continuously. For comparison, the result of a continuous tuning of the equilibrium by varying the length is shown in Fig. 21 by continuous curves. The solid curve corresponds to microtubule elongation. Two sample reverse branches of this function are also shown, which correspond to shortening from the two sample reversal points. It can be observed that the equilibrium centrosome position as a function of the microtubule length exhibits hysteresis.

Figure 22 illustrates the underlying cause: Elongation of microtubules expands the range of unstressed microtubule directions for which metastable forms no longer exist. The metastable forms lost during elongation are not restored during shortening, and this affects the equilibrium position of the centrosome. It does not return to the same position that it had when the length had the same value during elongation. As shown in Fig. 23, the equilibrium cytoskeleton as a whole exhibits memory of past variations of the microtubule length.

Considering these results in toto, one may observe that although the centrosome position can be controlled approximately linearly and approximately reversibly by varying the microtubule length, the evolution of the overall cytoskeleton form displays irreversibility. The conclusion that the evolution of the microtubule aster is in

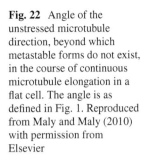

Fig. 22 Angle of the unstressed microtubule direction, beyond which metastable forms do not exist, in the course of continuous microtubule elongation in a flat cell. The angle is as defined in Fig. 1. Reproduced from Maly and Maly (2010) with permission from Elsevier

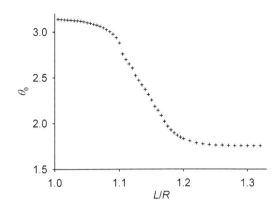

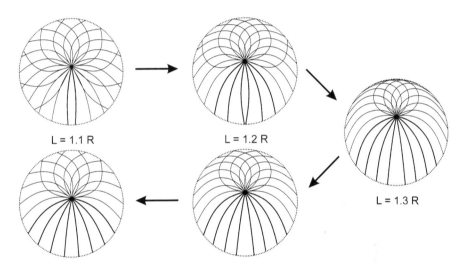

Fig. 23 Equilibrium forms of a microtubule aster in a flat cell during elongation and shortening of microtubules. Microtubule length is indicated in relation to the cell radius. Reproduced from Maly and Maly (2010) with permission from Elsevier

general irreversible should apply directly to more complex, realistic cytoskeletons. It does not depend on the simplicity of the aster structure in the computational examples shown, but only on the presence of metastable microtubule forms, which model feature is realistic, as discussed in the previous section.

The model of Maly and Maly (2010) establishes a qualitative difference between the flat and three-dimensional cases. In three dimensions, the aster adopts the "combed" conformation, in which all microtubules are buckled in the same direction (Fig. 3). Their distal (plus) ends are bent away from the direction of the centrosome displacement from the cell center. This kind of aster structure is "memoryless": The directions of microtubule buckling remain the same if the centrosome is moved by external forces or if the microtubule length changes, and there is only one equilibrium centrosome position for the given microtubule length. In contrast, the microtubules in the two-dimensional (flat) equilibrium aster can be bent in both directions from the unstrained direction of their emanation from the centrosome (Figs. 18 and 19). This gives the aster an "uncombed" appearance and is at the source of the memory effects exhibited by such an aster. Its structure depends on past perturbations, and the equilibrium position of the centrosome depends on the history of the microtubule length changes.

Another theoretical demonstration of irreversibility is provided by the model (Kim and Maly 2009) for the reorientation of the cell body in cytotoxic lymphocytes toward target cells engaged sequentially. The ability of lymphocytes to destroy sequentially engaged targets relies on reorientation of the secretory apparatus that directs secretion of cytotoxic granules and is associated with the centrosome and the microtubules (Valitutti et al. 1996; Depoil et al. 2005). From the experimental observations by Kuhn and Poenie (2002) of the movement and structure in the cytotoxic T lymphocytes, Kim and Maly (2009) derive the following idealizations on which they base their numerical model. The cell boundary has a constant shape that consists of an unattached round part and a flat part which is attached to the target cell. The flat part is referred to as the synapse, and is said to lie in the synaptic plane. The large rigid nucleus is coupled to the aster of microtubules converging near its surface. Together they comprise the cell body in this comparatively more complete model of the cellular structure. Like in the models considered heretofore, the cell boundary constrains the cell body, causing its deformation and limiting its mobility. Additionally, microtubules in this more complete model slide actively along the cell outline in the areas of contact with the targets. The movements of the cell body that are caused by the active sliding are opposed by microtubule bending elasticity and by viscous drag. The physical specification on the active sliding forces idealizes what should happen when microtubules come in contact with the cell boundary on whose inner surface (cell cortex) dynein molecules are anchored (Kuhn and Poenie 2002; Combs et al. 2006). It is assumed that the unit length of the contacting part of the microtubule will experience a constant tangential force directed to the distal end. The boundary in the synapse area is thus characterized by a certain force density that is by itself nondirectional. The magnitude of the force experienced by microtubules depends linearly on the length of the segment in contact with the boundary, and the direction of the force is determined entirely by the orientation of the said segment.

Figure 24 shows a numerical simulation in which the centrosome is initially oriented at 90° to the developing cell–cell interface. This angle has the highest likelihood if the spherical T cell comes in contact with the target surface in a random orientation. The model reproduces the experimental observations (e.g., Kuhn and Poenie 2002) that the centrosome becomes reoriented to the interface. Interestingly, stabilization of the centrosome orientation in the model is soon followed by development of pulse-like oscillations in its position (Fig. 24). These will be discussed in the next section. The prediction that concerns the subject of the present section is that the long-range reorientation results in an arrangement of microtubules that is very asymmetrical. On the side of the microtubule aster that was leading in the

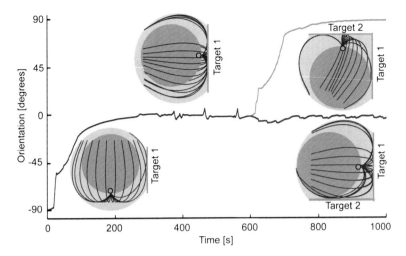

Fig. 24 Dynamics of the centrosome orientation in a lymphocyte developing sequentially two immunological synapses. *Insets* show the model cell, nucleus, centrosome, microtubules, and cell contact areas. Two alternative simulations differing in the position of the second synapse follow the simulation with the first synapse. Reproduced from Kim and Maly (2009) under the Creative Commons Attribution License

reorientation movement (the side next to which the synapse initially developed), a relatively tight "bundle" of microtubules is formed, which is separated from the microtubules that were trailing by a distinctive gap.

It can be observed parenthetically that "bundling" of microtubules is frequently seen also in the energy-minimization model discussed in the last section. They arise from multiple microtubules adopting similar conformations that are favored by bending in the confined intracellular space. Assumption of "cross-linking" that is common among the interpretations of microtubule bundling as observed in experiments has an obvious simpler alternative in the light of these calculations. Returning to the lymphocyte simulation, the predicted gap between the leading "bundle" and the trailing microtubules is indeed visible in a three-dimensional experimental image of an early T cell–target cell conjugate (figure 6a in Kuhn and Poenie 2002).

The strong asymmetry thus induced in the aster by the active sliding appears to be responsible for the ratchet-like behavior of the microtubule cytoskeleton, which is predicted by the Kim–Maly model when the T cell develops a second synapse. The centrosome readily reorients by another 90° in the same direction as it did the first time, but does not reorient in the opposite direction (Fig. 24). This easily testable prediction implies that the systems biomechanics of the cell body imposes specific constraints on the functionality of the cytotoxic lymphocytes during the immune response. More broadly, the emergent irreversibility arising from constraining of the cell body by the cell boundary deserves a detailed theoretical investigation in the context of various cell-biological phenomena and must be taken into account when designing and interpreting experiments.

Dissipative Oscillations

The previous sections have considered instability of trivial equilibria without regard to time, which gives rise to the distinctively complex and purposeful biological structures. We have also considered the irreversibility that emerges on the cellular level and imposes directionality on the development of the cellular structures with time. In this section, temporal instability will be treated. In general, it may be argued, instability of the latter kind gives rise to the perpetual nonrandom movement that is perhaps the best known attribute of living matter. Within the scope of this book we will be concerned here with the temporal instability that arises from the system-level mechanics of the interaction of the cell body and boundary. The cyclic motion of the cell body and boundary that is the cycle of cell divisions is the central mechanical phenomenon on which rests the existence of all advanced life forms. Considerable progress has been made in the systems-biomechanical approach to simpler cyclic motions of the cell body alone, which will be considered in this section.

Before embarking on this special subject, as permitted by today's state of knowledge, additional general reasons for studying the emergent periodic cell motion may be mentioned. Experimental observations of periodic movement have the potential to furnish dynamic data of higher quality for comparison with quantitative theory, because measurements in this case are effectively repeated on the same cell under the same conditions, eliminating the main sources of variability in quantitative cell research. Further, it is conceivable that the modes of oscillatory movement are the same as the modes of movement that appears non-oscillatory. Decomposition of non-periodic movement into periodic modes may lead to deterministic mechanical explanation for the complex movement that is the hallmark of cellular activity as observed under the microscope.

The cell body is known to oscillate within the cell boundary in various physiological contexts. An example of cyclic but complex, apparently multiperiodic and three-dimensional movement of the cell body within the confines of the cell boundary is presented by cytotoxic T lymphocytes of the immune system. These cells were the subject of an irreversibility model discussed in the last section and will be further treated here in regard to their periodic movements. In experiments, T lymphocytes display oscillations of their microtubule aster and of the mechanically coupled nucleus and membranous organelles. The cell body comprised of these linked structures oscillates within the lymphocyte's largely unchanging boundary next to and between the interfaces that the boundary has formed with target cells (Kuhn and Poenie 2002). The oscillations next to one target may facilitate the extrusion of the cytotoxic granules at that target, and the oscillations between interfaces appear to be the nonintuitive way in which the T cell achieves reorientation of its secretory apparatus between the simultaneously engaged targets.

A different example is provided by neuroblasts in the developing brain. In the pseudostratified layer of neural progenitors, nuclei undergo a cyclical motion which is coupled with the cell division cycle (Sidman et al. 1959). The orientation of these cells' cleavage plane varies, and has been linked to the alternatively proliferative and differentiative division, the latter producing daughter cells that become postmitotic neurons (Chenn and McConnell 1995). The apparently aperiodic tumble of these cells' mitotic spindles that precedes the selection of the cleavage plane orientation is well documented (Haydar et al. 2003). Neuroblasts are among the first candidates for further application of the systems-biomechanical models developed for the simpler and more regular oscillatory movement.

The spectrum of potential applications to movements of the cell body within the cell boundary that are not known to be periodic is further illustrated by the problem of distribution of nuclei in fungi. Upon germination of conidia in *Aspergillus nidulans*, for example, the nuclei divide and distribute along the germ tube in a microtubule-dependent and dynein-dependent fashion (Oakley and Morris 1980; Xiang et al. 1994).

Returning to the review of comparatively regular oscillatory movements, oscillations of the nucleus during meiosis in the fission yeast *Schizosaccharomyces pombe* have a period of approximately 10 min (Chikashige et al. 1994). The movement depends on dynein (Yamamoto et al. 1999). In its absence brought about by genetic interference with dynein, alignment of homologous chromosomes and recombination are suppressed (Yamamoto et al. 1999). The cited works speculate that the movement facilitates chromosome disentanglement.

A further class of objects for the periodic models is represented by fertilized eggs of various invertebrates. Their first division is accompanied by oscillations of the mitotic spindle, which appear to be a mechanism of search for the morphogenetically correct spindle orientation. So, for example, Dan and Inoué (2008) describe the preparation for the first division in the Atlantic surf clam *Spisula solidissima*. As one semiaster assumes the central position in the egg, the other is pushed out to the periphery, and its rays are bent flat (cf. the symmetry instability model in Fig. 13). Oscillations of this aster ensue, during which the spindle is pivoted at the position of the central aster. The oscillations appear to be a mechanism for searching for the morphogenetically correct orientation of the spindle, in which it stabilizes before division.

The flattening of the "rocking" aster in this well-described case is analogous to the flattening of the microtubules against the synapse in the cytotoxic T lymphocytes (Kuhn and Poenie 2002). In the lymphocytes, accumulation of the microtubule motor dynein on the cell cortex in the area of the synapse has been described (Combs et al. 2006). The rocking motion of the spindle during the second round of divisions in *Caenorhabditis elegans* appears to belong to the same class: The peripheral semiaster oscillates about the morphogenetically correct position, its flattening against the flattened cell–cell contact area is observed, and the dynein adapter dynactin is found on the cortex in the contact area (Skop and White 1998). These observations place special demands on the theoretical explanation of this apparently widely represented class of oscillatory motion of the cell body. It must consider large microtubule bends and dynein motor action on microtubules contacting the cortex tangentially.

The modern understanding of the periodic movement of the cell body can be traced conceptually to the earlier work on muscle and flagella oscillations. Pringle (1949) proposed that insect flight muscles were activated by stretch, and that this leads to autonomous oscillations in the antagonistic muscle pairs. The independence of this emergent mechanical behavior from the control by the nervous system provides, on the organ level, an analogy to the subject of systems biomechanics of the cell. Subsequently Brokaw (1975) showed theoretically that if disengagement of cross-bridges increased with the load, then coupled antagonistic pairs could develop and sustain oscillations. Indeed, the load-induced disengagement introduces the feedback that further weakens the losing side in the tug of war. This principle was applicable equally to myosin cross-bridges in muscle and dynein cross-bridges in flagella, and equally to pairs of flight muscles and diametrically opposite microtubule doublets in a flagellum. It mechanistically and quantitatively resolved the problem of the time delay requirement, which was implicit in Pringle's proposal.

The idea of the load-induced disengagement of molecular motors was applied to oscillations of the cell body by Grill et al. (2005). Their one-dimensional model was developed for the motion of the posterior spindle pole during the first division in *C. elegans*. Force generators exhibiting the stretch-induced disengagement were postulated to act on astral microtubules reaching the opposite sides of the cell boundary orthogonally to the spindle axis. The restoring force necessary for the oscillations was assumed to arise from the bending elasticity of the microtubules, which was modeled as amounting collectively to a simple Hookean spring tying the centrosome to the central position. The pulling strength must be sufficiently strong in this model relative to the strength of the restoring spring for the oscillations to arise and to be sustained. This prediction was further elaborated theoretically and tested experimentally (Pecreaux et al. 2006).

Subsequent work of Kozlowski et al. (2007) put the load-dependent force generators in a theoretical framework that included explicitly computed deformations of the astral microtubules in *Caenorhabditis*. The new model resolved mechanistically the problem of the physical conditions of the microtubule contact with the pulling cortex, which was implicit in the previous model. Kozlowski et al. demonstrated experimentally that the microtubules in this experimental system transition to disassembly upon contact with the cell boundary, and showed theoretically that the oscillatory dynamics can be sustained despite the brevity of each individual contact. As the authors note, the model was constructed to be representative of the conditions during anaphase B, when the oscillations take place in the *Caenorhabditis* egg, and would not be correct for the telophase, when the microtubules emanating from the former spindle pole become greatly deformed against the boundary of the posterior daughter cell.

The conditions of the very large deformations that are established at the end of the first division prevail also during the second division in *Caenorhabditis*, when the spindle pole oscillations are again observed (Keating and White 1998). They similarly prevail in lymphocytes conjugated with target cells (Kuhn and Poenie 2002), where the single interphase centrosome oscillates in close proximity to the cell–cell interface. [Structural and biomechanical similarities of these two systems have been reviewed in detail before (Maly 2011)]. The difficulty in applying the previously

Fig. 25 Oscillations of the lymphocyte centrosome at the immunological synapse. (a) Model cell structure plotted as in Fig. 24. (b) Oscillating microtubule system shown in projection onto the synaptic plane. The parts that are in contact with the synaptic surface and are experiencing the pulling are highlighted. Reproduced from Kim and Maly (2009) under the Creative Commons Attribution License

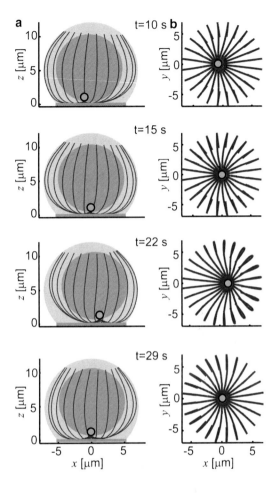

developed theoretical concepts led Kim and Maly (2009) to develop a new model. Elimination of the load-induced disengagement as an a-priori assumption led to the demonstration of its effective emergence on the cellular level from the large deformations of microtubules against the cytoplasmic viscous drag.

The construction of the Kim–Maly model for the cell-body movement in lymphocytes was discussed in the previous section. In the simulation of large-angle reorientation of the cell body (Fig. 24), the emergence of pulse-like oscillations could already be noticed. By analyzing further the mechanism of this instability, it was found that oscillations develop in the model even if the synapse is formed next to the initial location of the centrosome. Engagement of the microtubules with the pulling surface causes the model centrosome to move past (or away from) the center point of the interface. Although this point is the position of symmetry for the centrosome, the complete symmetry of the system in general does not develop, or cannot be sustained, in this dynamic model. The centrifugal movement of the centrosome eventually stops, and it begins the reverse motion, again approaching the center point and again overshooting (Fig. 25). The oscillations persist without noticeable

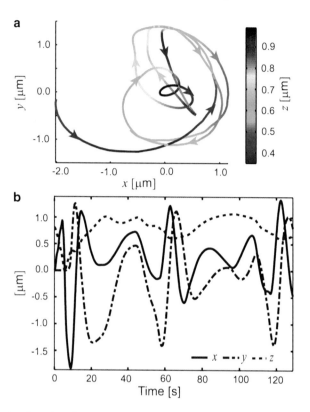

Fig. 26 Typical trajectories of centrosomes oscillating within a synaptic area. Coordinates as in Fig. 25. Note x–y phase shift leading to gyrations, and apparent beats. Reproduced from Kim and Maly (2009) under the Creative Commons Attribution License

systematic changes over at least 1 h of simulated physical time. Typically it appears that there are overlapping and interfering periodic motions. Also, oscillatory movements that are mostly tangential to the model cell–cell interface occur simultaneously with oscillatory movements that are orthogonal to it. Gyrations (looping motions parallel to the interface) can also be discerned in the complex trajectory of the model centrosome (Fig. 26).

Considering the origin of the oscillations and of the repeated overshooting exhibited by the centrosome, it is important to point out that inertia plays no role in intracellular movements due to the prevailing near-zero Reynolds number conditions. In fact, like in models for comparable types of intracellular movements (e.g., Cytrynbaum et al. 2003; Grill et al. 2005; Kozlowski et al. 2007), there is no mass in the Kim–Maly model. Also, the model is strictly deterministic, and therefore the deflections from the middle position of the centrosome are not due to molecular stochasticity.

Close inspection of the model predictions reveals that when the centrosome passes the middle point during oscillations, the microtubule aster shows significant asymmetry. This asymmetry is reversed next time the centrosome passes the middle point (Fig. 25b). Moreover, the microtubules are engaged with the pulling surface more to one side of the centrosome than to the other. The other side of the aster becomes engaged during the reverse swing (Fig. 25b). Similarly to the Kozlowski et al. model for the pronucleus oscillations in worm eggs, it can be observed that the distal ends of microtubules in the Kim–Maly model do not move appreciably during the oscillation cycle. This should be attributed to the cytoplasm viscosity dampening propagation of the elastic perturbation along the microtubules from their proximal parts which may be pulled by the cortex and which are coupled to the moving centrosome. As a result, when microtubules on one side are pulled and the centrosome shifts, the proximal parts of microtubules on the opposite side become lifted off the synaptic surface (Fig. 25). This makes the tug of war nonlinear: Whenever one side is winning, this weakens the opposing side. The temporal instability can be ascribed to this effect. Importantly for the cyclical nature of the movement that emerges from the instability, its range is limited by the deformation of the microtubules on the winning side. Their distal parts are bent against the side of the cell, and because of this the zone where they can contact the pulling surface cannot extend to the edge of the flat synaptic zone. Movement toward the edge therefore diminishes the pulling force. This gives the elastic relaxation of the trailing microtubules time to catch up and bring their proximal parts in apposition with the pulling surface. At this point, the microtubules that trailed are lying relatively flat on the synapse and are therefore experiencing a greater pulling force than the microtubules that led and are now contacting the synapse only with their highly curved parts. Movement in the reverse direction ensues (Fig. 25).

From the thermodynamic standpoint, the dissipative oscillations are driven by the non-potential forces of pulling. They arise from the energy-dissipating biochemistry of the dynein motors, which is coupled to the nonequilibrium metabolism of the living cell. In the course of the oscillations these non-potential active forces work against the similarly non-potential forces of viscous drag. At the same time, the mechanics of the described hysteretic loop is orchestrated by the potential forces of microtubule elasticity and by the heritable structural constraints of the cell body and boundary.

Simulations with different pulling force densities showed (Kim and Maly 2009) that the basic frequency of the oscillations is fairly insensitive to this parameter. Yet the overall pattern of oscillations changes abruptly when the pulling force density crosses a certain value (Fig. 27). Below approximately 140 pN/μm, the oscillations appear multiperiodic and continuous (Fig. 27a). Above approximately 150 pN/μm, the oscillations are pulse-like (Fig. 27c). In the relatively narrow range between approximately 140 and 150 pN/μm, the oscillations are continuous and pure, i.e., they exhibit a single frequency and amplitude. Only in this narrow intermediate range does the distance of the centrosome from the synaptic plane not oscillate (Fig. 27b).

The experimental estimates of the force exerted by a single cytoplasmic dynein molecule interacting with a microtubule, on the order of pN (Ashkin et al. 1990), limit the range of the pulling force densities that are of interest to 20–200 pN/μm. Below this range, there would be only a few molecular motors pulling on a given microtubule, giving rise to stochasticity. Above this range the number of motors

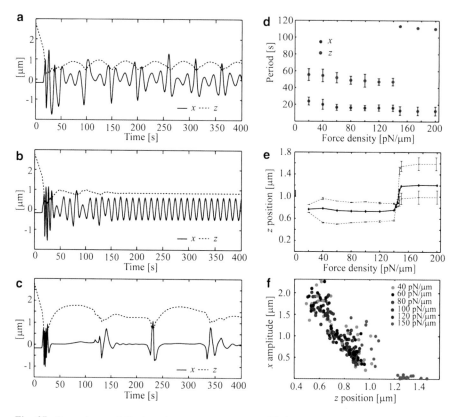

Fig. 27 Dependence of the lymphocyte centrosome oscillations on the pulling force density f. (**a**) $f = 100$ pN/μm. (**b**) $f = 143$ pN/μm. (**c**) $f = 200$ pN/μm. (**d**) Period of oscillations parallel and perpendicular to the synapse. The error bars are S.E. (**e**) The mean (*solid line*) and the range (*dashed lines*) of the centrosome distance from the synapse. The error bars are S.E. (**f**) Peak x deviation (amplitude) achieved at the given distance from the synapse (z). Reproduced from Kim and Maly (2009) under the Creative Commons Attribution License

would reach into the hundreds, which may not be realistic. The model showed that the period of oscillations parallel to the synapse remains near 15–20 s throughout the said force density range (Fig. 27d). This is close to the typical frequency seen in the experimental videos (Kuhn and Poenie 2002). While the numerical match is due to the assumed value of the effective viscosity of the cytoplasm (Kim and Maly 2009), the constancy of the period appears to be an intrinsic property of the mechanism modeled, because the viscosity value is the same in all simulations.

This intrinsic frequency of oscillations parallel to the synapse (x direction) is seen in its pure form when the orthogonal (z direction) oscillations are absent between 140 and 150 pN/μm (Fig. 27b). In the other two regimes (Fig. 27a, c), however, measurements show that the dominant x-frequency is approximately the same (Fig. 27d). The period of the z-oscillations is also mostly insensitive to the force density, except that it is much higher for all values above 150 pN/μm than it is below 140 pN/μm (Fig. 27d). In the intervening range, the z-oscillations are not sustained (Fig. 27b).

The range of the distance (z) of the centrosome from the synapse exhibits a similar step-like dependence on the pulling force density, collapsing fully in the narrow transition zone (Fig. 27e). It can be observed that the farther away from the synapse the centrosome is at any given time, the smaller the amplitude of the movement parallel to the synapse will be. Figure 27f shows that this relation is essentially independent of the force density and is quasi-linear. The exception to its linearity is, apparently, the natural limit of zero amplitude. When this limit is reached (this can happen only at high force densities), the amplitude–distance relationship displays a breakpoint at the axis intercept (Fig. 27f). The zero amplitude of motion parallel to the synapse is observed during the intervals between the pulses, as shown in Fig. 27c. Notably, the breakpoint of the x-amplitude versus z-position curve (Fig. 27f) is near the centrosome-synapse distance of 1 μm, the same distance that characterizes the breakpoints of the dependencies of the z-position and z-period on the force density (Fig. 27d, e).

Close examination shows that this transition corresponds in individual trajectories to complete but temporary loss of contact between the microtubule system as a whole and the synapse. This phenomenon appeared to arise from the viscous-drag-induced "liftoff" of the microtubules (Fig. 25a). During particularly vigorous movement that can occur at the higher force densities, not just one side but the entire microtubule system may lose contact with the synapse. In the absence of the active driving force, it will take the motile system a considerable time to relax elastically against the cytoplasmic drag and contact the synapse again. These periods of time correspond to the long, high arcs of the z-trajectory and no x-movement, as seen in Fig. 27c.

Overall, intuition does not keep up with the complexity of the movement arising in this relatively simply constructed mechanical system. At the same time, the complexity revealed by the sufficiently accurate numerical analysis closely resembles the multi-periodic and variable-amplitude movement seen in the experiments (Kuhn and Poenie 2002). It should be noted that the mechanistic explanation of this movement that is offered by the model will be difficult to test using the existing live-cell imaging techniques. Testing it would depend on resolving optically the small distances around the predicted breakpoints (~1 μm, Fig. 27d–f).

Following a simultaneous development of two synaptic areas, the model centrosome moves to one of them. After pausing at the first synapse (the pause can last for a significant period of time), the model centrosome spontaneously moves to the other synapse (Fig. 28a). The cycle of movement, pause, and movement to the other synapse appears to continue indefinitely with a well-defined periodicity. The characteristic delay before the reverse movement is as seen in the experiments (Kuhn and Poenie 2002). The model predicts that for the delay to take place, the angle between the two synaptic planes must be narrower than 150° (Fig. 28b). The angle was indeed sharp in the experiment (Kuhn and Poenie 2002). By adjusting the pulling force density and effective cytoplasm viscosity, both the duration of the pause and the

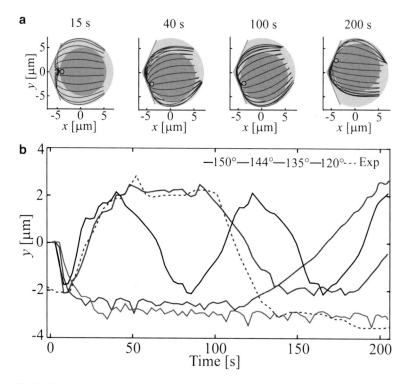

Fig. 28 Oscillations of the lymphocyte centrosome between two immunological synapses. (**a**) Model cell structure. (**b**) Trajectories of the centrosome predicted for the indicated angles between the synaptic planes. *Dashed curve* is an experimental trajectory. Reproduced from Kim and Maly (2009) under the Creative Commons Attribution License

duration of the movement phase could be reproduced (Fig. 28b). Whereas the match of the absolute model time scale to the experiment is data fitting and therefore not particularly significant, the fact that the computed phase of pause and the computed phase of migration can have the same relative duration as seen in the experiment is remarkable. It indicates that the simple deterministic mechanics may indeed account for the complex motility of the confined cell body in the conjugated T lymphocytes.

In the light of the Kim–Maly model, the pause of the centrosome and the associated secretory apparatus of the lymphocyte's cell body next to each of the synaptic areas of the cell boundary appears to arise from the delayed relaxation of microtubules that were trailing during the last period of cell body migration. This can be discerned by examining Fig. 28 closely, and it is the same factor that leads, in the extreme, to the irreversible, ratchet-like behavior of the model following very large reorientations (Fig. 24). In comparison, the medium-range migration between the two synapses can be reversible. Compared, on the other hand, with the relatively small-amplitude oscillations within a synapse (Fig. 25), the migration of the centrosome between the synapses winds the trailing microtubules around the nucleus more. It takes them longer to relax and contact the other synaptic area after the movement has been limited by the deformation of the previously leading microtubules. It is notable that the time that the secretory apparatus associated with the lymphocyte centrosome spends next to the given target is determined by the elasticity of the cytoskeleton. It is equally notable that the movement of the cell body to the other synapse is a mechanical consequence of its previous movement to the synapse where it is presently found.

In summary, purely deterministic and conceptually simple biomechanical models are capable of exhibiting complex, life-like cell body movements and deformations. The computational results demonstrate that the origin of the strikingly animate "wandering of aim" in T-killer cells of the immune system need not be sought in stochastic dynamics of individual molecules, or in indecision that might be exhibited by complex information processing in the T cell, or in indeterminate changes in the signaling input from the target cells. Instead, the rigorous numerical demonstration that a purely deterministic mechanical explanation exists for one of the most animate behaviors exhibited by cells suggests that similar explanations and supporting experimental evidence can be sought for other types of cell behavior that appear far from mechanistic.

Complex movements of the cell body within the cell boundary, such as those reviewed in the introduction to this section, are likely to have similar physical basis. Movements that are evidently oscillatory (albeit multiperiodic) can arise, in the light of the present theory, from the flexure of the confined cell body cytoskeleton and the tangential nature of forces exerted on it by the molecular motors of the boundary. Other complex movements which may not appear periodic may nonetheless be decomposed into a series of oscillations with different frequencies. More broadly, the relation of the cyclical movement of the confined cell body to one that is irreversible, as discussed in this section, and the asymmetric nature of the stable static cell body conformations, which was the subject of the previous sections, demonstrate intimate continuity of the most fundamental systems-biomechanical phenomena treated in this book.

It is hoped that the systems-biomechanical line of investigation into the emergence of life-like complex form and movement on the cellular level will be continued, as it has potential to illuminate the understudied role of sufficiently complex supramolecular mechanics in the physics of living matter. It is not inconceivable in the light of the preliminary results discussed here that the comparatively simple mechanical organization of the living matter on the cellular level, when viewed through the prism of adequately sophisticated quantitative systems theory, may prove to be causa prima for the complex form and movement that exemplify life. Indeed, the latter are mechanical phenomena (equilibria and disequilibria), and emerge on the cellular level. What had been lacking until recently was application of rigorous quantitative systems methods to cell mechanics. The maturing of the systems approach to biology in general, and the accessibility of adequate computing power to analyze models without simplification for the sake of computability have created the preconditions for a radically new methodology. Its application to cell biomechanics is opening the path to truly mechanistic understanding of the cellular form and movement, whose self-organization demarcates the living from the nonliving matter.

References

Arkhipov SN, Maly IV (2006a) Contribution of whole-cell optimization via cell body rolling to polarization of T cells. Phys Biol 3:209–219

Arkhipov SN, Maly IV (2006b) Quantitative analysis of the role of receptor recycling in T cell polarization. Biophys J 91:4306–4316

Arkhipov SN, Maly IV (2007) A model for the interplay of receptor recycling and receptor-mediated contact in T cells. PLoS One 2:e633

Arkhipov SN, Maly IV (2008) Retractile processes in T lymphocyte orientation on a stimulatory substrate: morphology and dynamics. Phys Biol 5:16006

Ashkin A, Schütze K, Dziedzic JM, Euteneuer U, Schliwa M (1990) Force generation of organelle transport measured in vivo by an infrared laser trap. Nature 348:346–348

Baluska F, Volkmann D, Barlow PW (2004) Cell bodies in a cage. Nature 428:371

Baratt A, Arkhipov SN, Maly IV (2008) An experimental and computational study of effects of microtubule stabilization on T-cell polarity. PLoS One 3:e3861

Bjerknes M (1986) Physical theory of the orientation of astral mitotic spindles. Science 234:1413–1416

Brangwynne CP, MacKintosh FC, Kumar S, Geisse NA, Talbot J, Mahadevan L, Parker KK, Ingber DE, Weitz DA (2006) Microtubules can bear enhanced compressive loads in living cells because of lateral reinforcement. J Cell Biol 173:733–741

Bray D (2001) Cell movements: from molecules to motility. Garland, New York

Brokaw CJ (1975) Molecular mechanism for oscillation in flagella and muscle. Proc Natl Acad Sci USA 72:3102–3106

Burakov A, Nadezhdina E, Slepchenko B, Rodionov V (2003) Centrosome positioning in interphase cells. J Cell Biol 162:963–969

Chenn A, McConnell SK (1995) Cleavage orientation and the asymmetric inheritance of Notch1 immunoreactivity in mammalian neurogenesis. Cell 82:831–841

Chikashige Y, Ding DQ, Funabiki H, Haraguchi T, Mashiko S, Yanagida M, Hiraoka Y (1994) Telomere-led premeiotic chromosome movement in fission yeast. Science 264:270–273

Combs J, Kim SJ, Tan S, Ligon LA, Holzbaur EL, Kuhn J, Poenie M (2006) Recruitment of dynein to the Jurkat immunological synapse. Proc Natl Acad Sci USA 103:14883–14888

Cytrynbaum EN, Scholey JM, Mogilner A (2003) A force balance model of early spindle pole separation in *Drosophila* embryos. Biophys J 84:757–769

Dan K, Inoué S (2008) Studies of unequal cleavage in molluscs II: asymmetric nature of the two asters. In: Inoué S (ed) Collected works of Shinya Inoué. World Scientific, Singapore

Depoil D, Zaru R, Guiraud M, Chauveau A, Harriague J, Bismuth G, Utzny C, Müller S, Valitutti S (2005) Immunological synapses are versatile structures enabling selective T cell polarization. Immunity 22:185–194

Dogterom M, Yurke B (1998) Microtubule dynamics and the positioning of microtubule organizing centers. Phys Rev Lett 81:485–488

Dupin I, Camand E, Etienne-Manneville S (2009) Classical cadherins control nucleus and centrosome position and cell polarity. J Cell Biol 185:779–786

Etienne-Manneville S, Hall A (2001) Integrin-mediated activation of Cdc42 controls cell polarity in migrating astrocytes through PKCzeta. Cell 106:489–498

Euteneuer U, Schliwa M (1992) Mechanism of centrosome positioning during the wound response in BSC-1 cells. J Cell Biol 116:1157–1166

Evans E, Yeung A (1989) Apparent viscosity and cortical tension of blood granulocytes determined by micropipet aspiration. Biophys J 56:151–160

Faivre-Moskalenko C, Dogterom M (2002) Dynamics of microtubule asters in microfabricated chambers: the role of catastrophes. Proc Natl Acad Sci USA 99:16788–16793

Farhadifar R, Röper JC, Aigouy B, Eaton S, Jülicher F (2007) The influence of cell mechanics, cell-cell interactions, and proliferation on epithelial packing. Curr Biol 17:2095–2104

Gliksman NR, Parsons SF, Salmon ED (1992) Okadaic acid induces interphase to mitotic-like microtubule dynamic instability by inactivating rescue. J Cell Biol 119:1271–1276

Gomes ER, Jani S, Gundersen GG (2005) Nuclear movement regulated by Cdc42, MRCK, myosin, and actin flow establishes MTOC polarization in migrating cells. Cell 121:451–463

Gotlieb AI, May LM, Subrahmanyan L, Kalnins VI (1981) Distribution of microtubule organizing centers in migrating sheets of endothelial cells. J Cell Biol 91:589–594

Grill SW, Gönczy P, Stelzer EH, Hyman AA (2001) Polarity controls forces governing asymmetric spindle positioning in the *Caenorhabditis elegans* embryo. Nature 409:630–633

Grill SW, Kruse K, Jülicher F (2005) Theory of mitotic spindle oscillations. Phys Rev Lett 94:108104

Gu B, Mai YW, Ru CQ (2009) Mechanics of microtubules modeled as orthotropic elastic shells with transverse shearing. Acta Mech 207:195–209

Harold FM (2001) The way of the cell. New York, Oxford

Haydar TF, Ang E Jr, Rakic P (2003) Mitotic spindle rotation and mode of cell division in the developing telencephalon. Proc Natl Acad Sci USA 100:2890–2895

Holy TE (1997) Physical aspects of the assembly and function of microtubules. Dissertation, Princeton University

Holy TE, Dogterom M, Yurke B, Leibler S (1997) Assembly and positioning of microtubule asters in microfabricated chambers. Proc Natl Acad Sci USA 94:6228–6231

Howard J (1998) Mechanics of motor proteins and the cytoskeleton. Sinauer, Cambridge

Howard J (2006) Elastic and damping forces generated by confined arrays of dynamic microtubules. Phys Biol 3:54–66

Hyman AA, White JG (1987) Determination of cell division axes in the early embryogenesis of *Caenorhabditis elegans*. J Cell Biol 105:2123–2135

Käfer J, Hayashi T, Marée AFM, Carthew RW, Graner F (2007) Cell adhesion and cortex contractility determine cell patterning in the *Drosophila* retina. Proc Natl Acad Sci USA 104:18549–18554

Keating HH, White JG (1998) Centrosome dynamics in early embryos of *Caenorhabditis elegans*. J Cell Sci 111:3027–3033

Kikumoto M, Kurachi M, Tosa V, Tashiro H (2006) Flexural rigidity of individual microtubules measured by a buckling force with optical traps. Biophys J 90:1687–1696

Kim MJ, Maly IV (2009) Deterministic mechanical model of T-killer cell polarization reproduces the wandering of aim between simultaneously engaged targets. PLoS Comput Biol 5:e1000260

Knoblich JA (2008) Mechanisms of asymmetric stem cell division. Cell 132:583–597

Koonce MP, Köhler J, Neujahr R, Schwartz JM, Tikhonenko I, Gerisch G (1999) Dynein motor regulation stabilizes interphase microtubule arrays and determines centrosome position. EMBO J 18:6786–6792

Kozlowski C, Srayko M, Nedelec F (2007) Cortical microtubule contacts position the spindle in *C. elegans* embryos. Cell 129:499–510

Kuhn JR, Poenie M (2002) Dynamic polarization of the microtubule cytoskeleton during CTL-mediated killing. Immunity 16:111–121

Kupfer A, Singer SJ (1989) Cell biology of cytotoxic and helper T cell functions: immunofluorescence microscopic studies of single cells and cell couples. Annu Rev Immunol 7:309–337

Kupfer A, Louvard D, Singer SJ (1982) Polarization of the Golgi apparatus and the microtubule-organizing center in cultured fibroblasts at the edge of an experimental wound. Proc Natl Acad Sci USA 79:2603–2607

Kupfer H, Monks CKF, Kupfer A (1994) Small splenic B cells that bind to antigen-specific T helper (Th) cells and face the site of cytokine production in the Th cells selectively proliferate: immunofluorescence microscopic studies of Th-B antigen-presenting cell interactions. J Exp Med 179:1507–1515

Levy JR, Holzbaur ELF (2008) Dynein drives nuclear rotation during forward progression of motile fibroblasts. J Cell Sci 121:3187–3195

Lillie F (1901) The organization of the egg of *Unio*, based on a study of its maturation, fertilization, and cleavage. J Morphol 17:227–292

Longabaugh WJR, Bolouri H (2006) Understanding the dynamic behaviour of genetic regulatory networks by functional decomposition. Curr Genomics 7:333–341

Lowin-Kropf B, Smith Shapiro V, Weiss A (1998) Cytoskeletal polarization of T cells is regulated by an immunoreceptor tyrosine-based activation motif-dependent mechanism. J Cell Biol 140:861–871

Maly IV (2009) Introduction: a practical guide to the systems approach in biology. In: Maly IV (ed) Systems biology. Springer, New York

Maly IV (2011) Systems biomechanics of centrosome positioning: a conserved complexity. Commun Integr Biol 4:230–235

Maly IV (2012) Equilibria of idealized confined astral microtubules and coupled spindle poles. PLoS One 7:e38921

Maly VI, Maly IV (2010) Symmetry, stability, and reversibility properties of idealized confined microtubule cytoskeletons. Biophys J 99:2831–2840

McCarthy EK, Goldstein B (2006) Asymmetric spindle positioning. Curr Opin Cell Biol 18:79–85

Mickey B, Howard J (1995) Rigidity of microtubules is increased by stabilizing agents. J Cell Biol 130:909–917

Nedelec F (2002) Computer simulations reveal motor properties generating stable anti-parallel microtubule interactions. J Cell Biol 158:1005–1015

Oakley BR, Morris NR (1980) Nuclear movement is β-tubulin-dependent in *Aspergillus nidulans*. Cell 19:255–262

Orth JD, Thiele I, Palsson BØ (2010) What is flux balance analysis? Nat Biotechnol 28:245–248

Pearson CG, Bloom K (2004) Dynamic microtubules lead the way for spindle positioning. Nat Rev Mol Cell Biol 5:481–492

Pecreaux J, Röper J-C, Kruse K, Jülicher F, Hyman AA, Grill SW, Howard J (2006) Spindle oscillations during asymmetric cell division require a threshold number of active cortical force generators. Curr Biol 16:2111–2122

Piel M, Meyer P, Khodjakov A, Rieder CL, Bornens M (2000) The respective contributions of the mother and daughter centrioles to centrosome activity and behavior in vertebrate cells. J Cell Biol 149:317–329

Pinot M, Chesnel F, Kubiak JZ, Arnal I, Nedelec FJ, Gueroui Z (2009) Effects of confinement on the self-organization of microtubules and motors. Curr Biol 19:954–960

Pringle JWS (1949) The excitation and contraction of the flight muscles of insects. J Physiol 108:226–232

Rubinstein B, Larripa K, Sommi P, Mogilner A (2009) The elasticity of motor-microtubule bundles and shape of the mitotic spindle. Phys Biol 6:016005

Schuh M, Ellenberg J (2008) A new model for asymmetric spindle positioning in mouse oocytes. Curr Biol 18:1986–1992

Sidman RL, Miale IL, Feder N (1959) Cell proliferation and migration in the primitive ependymal zone: an autoradiographic study of histogenesis in the nervous system. Exp Neurol 1:322–333

Siller KH, Doe CQ (2009) Spindle orientation during asymmetric cell division. Nat Cell Biol 11:365–374

Skop AR, White JG (1998) The dynactin complex is required for cleavage plane specification in early *Caenorhabditis elegans* embryos. Curr Biol 8:1110–1116

Stowers L, Yelon D, Berg LJ, Chant J (1995) Regulation of the polarization of T cells toward antigen-presenting cells by Ras-related GTPase CDC42. Proc Natl Acad Sci USA 92:5027–5031

Symes K, Weisblat DA (1992) An investigation of the specification of unequal cleavages in leech embryos. Dev Biol 150:203–218

Théry M, Racine V, Pépin A, Piel M, Chen Y, Sibarita JB, Bornens M (2005) The extracellular matrix guides the orientation of the cell division axis. Nat Cell Biol 7:947–953

Tran PT, Marsh L, Doye V, Inoué S, Chang F (2001) A mechanism for nuclear positioning in fission yeast based on microtubule pushing. J Cell Biol 153:397–412

Ueda M, Gräf R, MacWilliams HK, Schliwa M, Euteneuer U (1997) Centrosome positioning and directionality of cell movements. Proc Natl Acad Sci USA 94:9674–9678

Valitutti S, Muller S, Dessing M, Lanzavecchia A (1996) Different responses are elicited in cytotoxic T lymphocytes by different levels of T cell receptor occupancy. J Exp Med 183:1917–1921

Watt FM, Hogan BL (2000) Out of Eden: stem cells and their niches. Science 287:1427–1430

Whittaker JR (1980) Acetylcholinesterase development in extra cells caused by changing the distribution of myoplasm in ascidian embryos. J Embryol Exp Morphol 55:343–354

Xiang X, Beckwith SM, Morris NR (1994) Cytoplasmic dynein is involved in nuclear migration in *Aspergillus nidulans*. Proc Natl Acad Sci USA 91:2100–2104

Yamamoto A, West RR, McIntosh JR, Hiraoka Y (1999) A cytoplasmic dynein heavy chain is required for oscillatory nuclear movement of meiotic prophase and efficient meiotic recombination in fission yeast. J Cell Biol 145:1233–1249

Yvon AMC, Walker JW, Danowski B, Fagerstrom C, Khodjakov A, Wadsworth P (2002) Centrosome reorientation in wound-edge cells is cell type specific. Mol Biol Cell 13:1871–1880

Printed by Publishers' Graphics LLC
DBT130701.15.15.31